PRACTICAL BUILDING CONSERVATION

VOLUME 1

STONE MASONRY

Practical Building Conservation Series:

Volume 2: BRICK, TERRACOTTA AND EARTH
Volume 3: MORTARS, PLASTERS AND RENDERS
Volume 4: METALS
Volume 5: WOOD, GLASS AND RESINS

PRACTICAL BUILDING CONSERVATION

English Heritage Technical Handbook

VOLUME 1

STONE MASONRY

John Ashurst
Nicola Ashurst

Photographs by Nicola Ashurst
Graphics by Iain McCaig

Gower Technical Press

Published by
Gower Technical Press Ltd,
Gower House,
Croft Road,
Aldershot,
Hants GU11 3HR,
England

Reprinted 1989, 1990

British Library Cataloguing in Publication Data

Ashurst, John
Practical building conservation: English Heritage technical handbook.
Vol. 1 Stone Masonry
1. Historical buildings—Great Britain—Conservation and restoration
I. Title II. Ashurst, Nicola
720'.28'8 NA109.G7

ISBN 0 291 39745 X

Printed and bound in Great Britain by
BPCC Wheatons Ltd, Exeter

CONTENTS

FOREWORD

by Peter Rumble CB, Chief Executive, English Heritage

Over many years the staff of the Research, Technical and Advisory Service of English Heritage have built up expertise in the theory and practice of conserving buildings and the materials used in buildings. Their knowledge and advice have been given mainly in respect of individual buildings or particular materials. The time has come to bring that advice together in order to make available practical information on the essential business of conserving buildings – and doing so properly. The advice relates to most materials and techniques used in traditional building construction as well as methods of repairing, preserving and maintaining our historic buildings with a minimum loss of original fabric.

Although the five volumes which are being published are not intended as specifications for remedial work, we hope that they will be used widely by those who write, read or use such specifications. We expect to revise and enlarge upon some of the information in subsequent editions as well as introducing new subjects. Although our concern is with the past, we are keenly aware that building conservation is a modern and advancing science to which we intend, with our colleagues at home and abroad, to continue to contribute.

The Practical Building Conservation Series

The contents of the five volumes reflect the principal requests for information which are made to the Research, Technical and Advisory Services of English Heritage (RTAS) in London.

RTAS does not work in isolation; it has regular contact with colleagues in Europe, the Americas and Australia, primarily through ICOMOS, ICCROM, and APT. Much of the information is of direct interest to building conservation practitioners in these continents as well as their British counterparts.

English Heritage

English Heritage, the Historic Buildings and Monuments Commission for England came into existence on 1st April 1984, set up by the Government but independent of it. Its duties cover the whole of England and relate to ancient monuments, historic buildings, conservation areas, historic gardens and archaeology. The commission consists of a Chairman and up to sixteen other members. Commissioners are appointed by the Secretary of State for the Environment and are chosen for their very wide range of relevant experience and expertise. The Commission is assisted in its works by committees of people with reputation, knowledge and experience in different spheres. Two of the most important committees relate to ancient monuments and to historic buildings respectively. These committees carry on the traditions of the Ancient Monuments Board and the Historic Buildings Council, two bodies whose work has gained them national and international reputations. Other advisory committees assist on matters such as historic gardens, education, interpretation, publication, marketing and trading and provide independent expert advice.

The Commission has a staff of over 1,000, most of whom had been serving in the Department of the Environment. They include archaeologists, architects, artists, conservators, craftsmen, draughtsmen, engineers, historians and scientists.

In short, the Commission is a body of highly skilled and dedicated people who are concerned with protecting and preserving the architectural and archaeological heritage of England, making it better known, more informative and more enjoyable to the public.

ACKNOWLEDGEMENTS

The authors gratefully acknowledge the assistance of Dr Clifford Price, Head of the Ancient Monuments Laboratory, English Heritage, in the reading of the texts.

1 REPAIR AND MAINTENANCE OF STONE*

1.1 RECOGNITION AND DIAGNOSIS OF PROBLEMS

The repair and maintenance of stone buildings, whether complete or in a ruined condition, are amongst the most important building conservation activities. If the repairs are thoughtfully and competently carried out, the life of a stone structure may be extended indefinitely with dignity; if repaired in ignorant and unskilled ways unnecessary destruction and disfigurement will take place.

Architects and others responsible for the repair of old masonry buildings should be able to recognize and diagnose problems, to know where the right replacement materials may be obtained and to know where the appropriate skills may be found to carry out the work. There is no insuperable problem in the UK relating to stone supply, either the original or an acceptable substitute, nor is there any lack of expertise in working and fixing stone, although a persistent national training programme must be enlarged and maintained. This volume describes some of the more common problems of stone and the basic work associated with replacement stones.

The problems of stone decay are complex and have been extensively studied internationally for over half a century. The select bibliography at the end of this chapter gives the most concise and available sources of information which are likely to be of use to the practising architect. (See also Volume 4, Chapter 12 'Technical bibliography').

There are many ubiquitous symptoms of weathering, deterioration and decay. The effects and causes may be listed in two categories. The first category can be described as effects and causes associated with construction, detailing and use. In other words, there is a human element involved.

*Extensive references are made in this chapter to *The Conservation of Building and Decorative Stone* Ashurst J, Dimes F G and Honeyborne D B to be published by Butterworth Scientific 1988

Decay category 1: construction, detailing, use

a Cracking of stones and joints due to structural movement and settlement of large areas of a building, or unequal settlement of elements tied to each other

b Cracking due to poor detailing and construction, such as provision of inadequate bearings for lintels or thin stone facings to poor quality core filling or cills cracking over hard spots in bedding

c Spalling, splitting and lifting due to the volume increase of rusting, embedded iron cramps, straps, window or door ferramenta

d Staining and decay due to inadequate protection of wall heads, copings, cornices and other projections

e Staining, decay and open joints with lime runs due to neglect of joint condition and free water access into the wall

f Water staining and scouring sometimes associated with frost damage due to inadequate provision for rainwater disposal or failure of rainwater goods

g Cracking and advanced decay around mortar which is too dense and impervious for the stone

h Spalling and other damage around joints caused by careless cutting out to re-point

i Pitting and dishing from careless air abrasive or disc cleaning

j Staining and efflorescence associated with inexpert or inappropriate chemical cleaning

k Surface discolouration, flaking and pitting due to shallow surface 'preservative' treatments. To this list might be added decay due to poor selection or misuse of stone. The classic example of the latter is the placing of limestone over sandstone in a building, resulting in the accelerated decay of the sandstone.

These examples, however, and others in category 1 overlap with the second category. Under category 2 can be listed the effects and causes of loss of stone due to attack by acid gases in the air, frost action and salt crystallization. Of these three, the last is the most damaging and the most universal.

Decay category 2: weathering

a Roughened surfaces on limestone, marble and calcareous sandstones where regularly washed and 'etched' by rain

b As (a) above but preferentially weathering out weaker areas such as soft sand pockets or clay beds

c As (a) above but coupled with substantial softening and spalling due to polluted rain and condensation attack on the calcite or dolomite binder in some limestones, magnesian limestones and calcareous or dolomitic sandstones

d Acid etching of marble and some limestones by the presence of acid-secreting lichens

e Cracking, splitting and spalling of surface of limestones in areas sheltered from direct rain-washing, due to the formation of crystalline sulphate skins and their subsequent failure

f Uniform thickness scales separating from sandstone and following the profile of the surface, described as 'contour scaling' and associated with wetting and drying cycles, migration of natural cementing matrices and surface pore blocking with materials deposited from the atmosphere

g Scaling, powdering surfaces of all stones associated with efflorescence and with examples (d) and (e) above, due to soluble salt crystallization damage

h Lens-shaped spalls or splitting or 'map-cracking' especially in copings, weatherings, plinths, paving and steps due to freezing of trapped water

i Softening and general deterioration of slate due to wetting and drying of slates containing calcite or unstable pyrite

j Distortion and buckling of marble due to stress release when fresh from the quarry and in thin slabs, or subsequently due to inadequate support of slabs coupled with wetting and drying, heating and cooling.

The problems with salts

Only example (e) needs to be further explained here, because it can take place in all stones, mortar and rendering irrespective of their chemical composition and relatively independently of their environment. A salt solution transferred to the pores of the stone will, when the water evaporates, deposit the salt(s) on the surface of the stone ('efflorescence') or within its pores ('cryptoflorescence') or in both. Repeated wetting and drying cycles, each leading to a re-dissolving and recrystallisation of the salt(s) exerts a pressure on the walls of the pores. When this pressure exceeds the internal strength of the stone there will be damage in the form of powdering and fragmentation.

The sources of these problematical salts are numerous and varied. Diagnosis of the cause of decay of stone will not, of course, cure the problem but it will provide a sound basis for deciding what remedial treatment is needed in the form of repair and maintenance.

1.2 ROUTINE MAINTENANCE

Sensible maintenance can sometimes reduce the necessity for repair and replacement, whether the building is a roofless monument or in full use. Examples of routine maintenance are keeping gutters, downpipes, roof coverings, flashings,

and hoppers in good working order, joints properly pointed, excessive vegetation removed or controlled, painting of ferrous metal and judicious cleaning.

Good maintenance anticipates problems and does not create new ones.

1.3 MAINTENANCE AND REPAIR OF RUINED MASONRY BUILDINGS

Ruined masonry buildings have particular problems which cannot necessarily be resolved using traditional masonry techniques. The stability of large ruined structures may be seriously affected by the loss of significant structural elements. The policy of the Ancient Monuments Works section (Historic Buildings and Monuments Comission for England, 'English Heritage', formerly under Department of the Environment and Ministry of Public Building and Works) has long been to 'consolidate as found', using the minimum of discreet intervention. Because of the archaeological and historical significance of such sites the work is always carried out by a team consisting of archaeologists, architects, engineers and a specialist workforce. These sites contain, potentially, so much information above and below ground that familiarity with teamwork procedures is essential, so that evidence is not missed or destroyed. Long years of experience are essential to take on much of the work of this kind correctly, which is why there is a strong tradition of using directly employed labour on historic conservation work. Inexperienced teams, however well versed in building trades techniques, can find great difficulty in adjusting to the attitude and techniques required on this work.

Walls below modern ground level

Excavated walls which are to remain exposed to view will require, as soon as possible, a programme of consolidation and repair followed by some plan for maintenance. Stones and mortar which have lain for centuries in saturated ground or dry sand may have survived in excellent condition due to these stable environments. Once exposed, however, they may begin to show signs of deterioration fairly quickly, as exposure to wind, sun and rain sets up wetting and drying cycles and the destructive crystallization of soluble salts begins. Winter conditions may bring the additional hazard of frost to walls saturated with water and substantial losses may be incurred in one night.

Should consolidation work be delayed, temporary protection must be provided, appropriate to the risks of exposure. Such protection may range from sand or straw under black polyethylene sheets tied and weighted down, to insulating quilts, temporary boxing filled with polystyrene beads, or to temporary scaffold frame structures which can double up as protection for the excavation or maintenance team and may even be heated.

Walls standing above ground level

If the full, experienced team described above is not available, it may still be essential that necessary first-aid is carried out to ensure that further collapse, disintegration or vandalism are kept to a minimum. Emergency work of this kind

Ruined masonry buildings present particular problems of repair and consolidation. This illustration shows consolidated core-work and facing stones now requiring further maintenance. New stones are only rarely introduced in a situation such as this, where the objective is to 'preserve as found'. Note that the core work, exposed through centuries of weathering and stone robbing is not brought forward of the line formed by the tails of the missing stones, nor is it consolidated in a way which would make it appear to be random rubble instead of fill.

may include the provision of secure fencing, formwork to support vaults and arches in danger of collapse and strutting and shoring against leaning and bulging walls. Wall head protection may also be necessary. Features of particular value may need to be protected by temporary roofs. Low, frost-susceptible walls may need the protection of straw-filled quilts.

Control of organic growth

Where walls stand above ground, control of woody weeds may be necessary. The root systems of many plants and trees can feed on wall core and disrupt stones. The ivy, hedera helix, is particularly dangerous because of its rapid growth and the searching effects of its aerial roots. Lichens and mosses may also need to be removed as part of a repair or maintenance programme.

Volume 1, Chapter 2 'Control of organic growth' deals with these aspects.

Basic consolidation: 'rough racking'

Ruined masonry buildings suffer from the exposure of elements to the weather which were once protected by roofs, capstones and facing stones; they also suffer from stone-robbing. Over the centuries the neglected masonry building has provided a tempting source of ready-cut and dressed stone which has been prised away from the face, leaving rough, core-filling exposed. Core is very vulnerable to water penetration and frost damage.

'Rough racking' is the term used for the general teatment of these exposed areas of core such as broken wall faces and wall tops; it is also used to describe rubble masonry introduced to support or strengthen overhanging masses which would otherwise hold water.

The aim in this work, which demands a high level of skill and experience, is to reproduce the appearance of existing exposed core and at the same time to provide adequate protection for the wall. Whilst the same care should be taken with rough racking as with the pointing of facework, the distinction between the two should always be in mind when carrying out the work and must always be made clear. Consolidated corework must not be finished to look like rubble facing. There will always be a much higher percentage of mortar exposed in core than in facing.

Should corework form part of any original openings such as doorways or arrow slits, necessary allowance must be made for the missing facework of these openings by keeping the core back sufficiently to allow for the space which the facework originally occupied.

When dealing with narrow walls, where facing stones are missing, it may be found necessary to bring the new corework out nearly to the wall face to obtain sufficient strength, stability and support for any overhanging features. In such cases, the impression of the back of the missing face stone should be formed in the corework. This can be achieved by building face stones in and subsequently withdrawing them before the mortar takes its final set.

In many cases, corework will contain poor weathering material, such as chalk and decayed lime mortar. If such material is retained and exposed to the action of the weather it is likely to disintegrate completely in a few winters. For reasons of maintenance and economy, therefore, it is usually desirable to replace lumps of chalk or disintegrating stone of poor weathering quality with stone rubble of a more durable kind. Substitute stones must always match the original in appearance as closely as possible.

Mortar mixes

The composition of the mortar is of considerable importance, and although the original aggregates should be as closely copied in types, sizes and distribution as possible, the mix may need to be modified to improve the weathering characteristics. Sometimes it is possible to reuse original aggregates in with the new mix if they are distinctive and difficult to replace, but it is usually possible to match grits, sands, sea shells and crushed brick which form the bulk of most historic mortars.

'Weathering' the mortar by brush stippling and/or spraying after the initial set is recommended, so that the new work is unobtrusive and allows maximum drying out from the wall. This technique should not be carried out to the extent of showing more aggregate than natural weathering would expose.

Mixes approximating 1:2:9 or 1:1:6 (cement:lime putty:aggregate) may be necessary on exposed wall tops, but the cement gauging should be reduced by $\frac{1}{2}$ or $\frac{2}{3}$ wherever possible, and especially where the core stone is only of moderate durability. If imported French hydraulic lime is used the mix may be 2:5 (hydraulic lime:aggregate) for all but the most severe exposures, when one of the cement-gauged mixes should be used. If cement gauging is used with lime in replacing a very white original, white cement should be used. The use of pigments for colouring core binder is not recommended, and every attempt should be made to achieve the original colour with selected aggregates or naturally coloured binders such as the French hydraulic lime, or a pozzolanic PFA, which will provide a light buff colour. (French hydraulic lime may at times be grey.)

The aim is always to preserve the broken outline of the ruined wall rather than impose regular levels and falls. To do this, grass and plants must be cut out, the stones cleaned off and, where loose or forming water traps, numbered, photographed, lifted and rebedded. All soil should be removed, so that it is sometimes necessary to excavate deeply into the wall head. Stones are replaced as found except for modifications needed to ensure that water does not pond on the surface.

In all cases careful study of the original material and adequate records of appearance should be made, so that it may be successfully reproduced and readily interpreted. Considerable experience is required to 'read' corework, which may contain shadowy evidence of, for instance, vault springers, beam bearings or alterations in the building. These can easily be overlooked and destroyed during consolidation and it is therefore critical that a professional archaeological survey precedes all work of this kind.

Where corework is found in reasonable condition it may be necessary only to take out loose joints, tamp and point and form proper falls to shed water as easily as possible from wall tops.

Structural intervention

'Overkill' response to distortion and fracturing in masonry buildings is frequently needless and sometimes dangerous. Treatment of symptoms, rather than causes, can result not only in unnecessary expense but also in compounding a problem or introducing entirely new dimensions to it. Engineers, just as much as other professional consultants, need particular experience of masonry structures to be able to diagnose and recommend correctly. 'Belt-and-braces' interventions which have more to do with professional insecurity than structural stability are to be strenuously avoided. This is not a recommendation to take chances with lives and safety but to ensure that adequate expertise is employed to make the right decisions.

Common problems, especially in ruined buildings, which may require structural intervention, may be listed as follows:

- Fracturing due to unequal settlement of building elements
- Local bulging due to loss of integrity of rubble cored walls
- Leaning due to settlement or loss of restraint
- Failures due to overloading (often the result of alterations or additions)
- Fractures due to loss of bearings.

Whether or not intervention is necessary may be obvious during one visit or may take many months or even years of careful monitoring. Whereas the introduction of new masonry on a take-down, rebuild and/or replace basis may be sound building practice, it is not desirable in historic structures and is seen as the last resort. To repair and consolidate in situ with minimum disturbance leaving minimum evidence of the intervention should be the objective.

Structural repairs such as wall head beams, fracture stitching, underpinning and secret lintels must only be used as the result of careful diagnosis of the problem. These structural interventions will often be accompanied by the need to grout. Grouting techniques are described in Chapter 3.

1.4 REPAIR AND REPLACEMENT OF STONES

Repair and replacement must always be preceded by a thorough survey which looks, firstly, at the whole condition of the walls, secondly at the condition of individual stones and thirdly, at the condition of the joints.

Where stones have become so badly decayed or damaged that some intervention is essential, the options may be summarized as follows:

1 Provide local protection in the form of flashings, weatherings or temporary shelters

2 Carry out minimal descaling and mortar filling (see Chapter 11)

3 Stitch and fill fractured stones

4 Carry out minimal piecing-in of stones

5 Carry out replacement of stones

6 Carry out (4) in mortar (plastic repair – see Chapters 4 and 8)

7 Provide a surface treatment, such as a consolidant (see Chapters 8, 9 and 10).

The treatment of joints is described in Volume 2, Chapter 4.

Criteria for identifying stones to be replaced

Value of the stones

The intrinsic value of any worked stone in a building clearly varies considerably with the age of the building and the quality and condition of the detail. The replacement of ashlars, if essential, will certainly be less controversial than the replacement of medieval detail, but no replacement is to be undertaken lightly. Good masonry practice is not always good conservation practice.

To determine 'value' is not a simple matter and it is difficult to make rules about it. Perhaps it is sufficient to say that copies should not usually be attempted of carving too distant from us in time and culture. Sometimes the value of individual stones, especially in Renaissance and later work, is subordinate to the value of the architectural design of the building. The line of a string with its important, unbroken shadow may be considered of far more importance than the preservation of a few decayed stones in its length.

Function of the stones

The function of any stone which is under consideration for replacement must be clearly understood. Decaying stones which have a structural role to play and on which the stability and survival of other stones or other elements of the structure depend have a clear priority for replacement almost regardless of their intrinsic value. Typical stones in this category are copings, weatherings, quoinstones and voussoirs.

Timing of the replacement

The expense of a scaffolding is, in itself, an encouragement to replace 'border-line' stones which might, or might not survive until the next scaffolding programme.

The estimated life of such stones must depend entirely on the experience of the architect and his masons who should use their knowledge to balance their concern for the building with the need to preserve for posterity as much original fabric as possible.

Alternative remedial work

Alternatives to removing stone must always be considered first. Such measures may simply involve attention to open joints or the provision of a lead flashing or discreet gutter over a label mould and stop, or the removal of an impermeable cement pointing, or a surface treatment designed to protect with a sacrificial

In a historic building only an essential minimum of new stone replacement should be carried out. In the heavily weathered sandstone of this medieval buttress the new stones may appear strange because of their relative sharpness and clean appearance, but they are geologically correct and in time will weather back to look more 'at home' in the masonry. Attempts to 'distress' the new stones or set them to lines other than the correct original should be strenuously avoided.

layer or deeply penetrating consolidant. In this category, too, may be the design and provision of a protective screen or roof over, for example, a rood or tympanum.

Protection in the form of lead dressings is a much simpler expedient which may be introduced discreetly to assist stone elements with a particularly difficult weathering job to do. Thus a code 4 flashing may be dressed over a small string, label mould or transom neatly fixed into a carefully prepared chase, or in ruined buildings a code 5 lead cap may, for instance be dressed over the exposed top of a tas-de-charge. The unique ability of lead to take up through careful bossing, the informal, soft outlines of weathered or damaged stone is particularly useful. Modest expenditure on protective flashing and capping may secure much valuable detail if thoughtfully placed and skilfully executed. Partially damaged strings or other projections may sometimes be built up in mortar with minimum disturbance if a lead flashing is subsequently run along the top. Fixing and detailing is, however, critical to performance and appearance. The method of securing lead into chases of joints recommended by the Lead Development Association is illustrated in Volume 4, Chapter 9.

Cutting out old stone

Once decisions have been made, based on the above criteria, on which stones are to be replaced, these will need to be indicated on a record drawing or photograph or, ideally, on a photogrammetric survey drawing. They must also be clearly marked on site with an indelible marker. Once a decision has been taken on replacement the most economical and sensible way of carrying out the work must be determined. In general, new stones will need to be 100 mm (4 in) on bed unless the stones are very small or only local piecing in of a larger stone is taking place, but it is often cheaper to remove an old stone completely than to face it with a new 100 mm (4 in) skin. During the marking up procedure notes should also be prepared for the specification of necessary temporary supports which may simply be wooden plates, struts and folding wedges or, when lintels, arches and vaults are involved, full centring.

The physical process of cutting out the old stone will vary according to the situation, but due care is required to ensure that the surviving stones adjacent are not damaged. Cutting of perimeter joints may be carried out with a masonry saw or a diamond cutting disc mounted on a power tool. If the old stone is to be retained the cut will first be made by diamond disc (or a purpose-made fine saw blade, in the case of a fine joint and hard mortar), or with a quirk or plugging chisel in the case of a wide joint.

If the stone is to be wasted it may be broken down with vertical saw cuts after the initial cutting or broken up with a hammer and chisel. Smaller-scale piecing in or indenting will involve cutting into an existing stone to remove a pocket of decay. Piecings may be very small indeed in good-quality work (20 mm square on face, for example) and the cut out will be made with small, sharp chisels and small saw blades to a neat, square profile.

Large stones may be 150 mm (6 in) on bed, or more for bonders whose tails are to be set into corework. If a large area is to be faced up with new stone it is

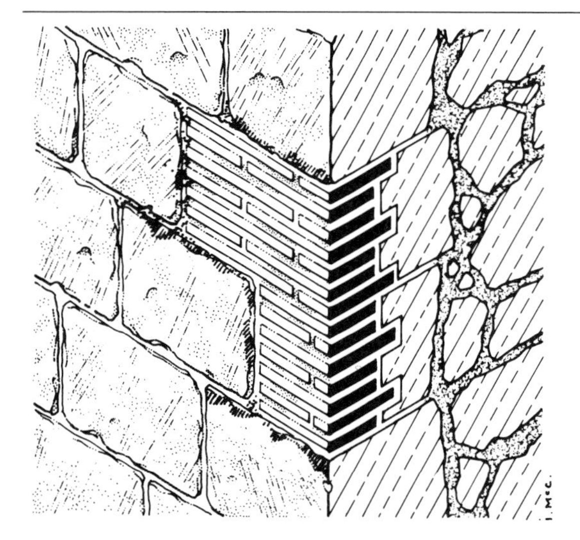

Indenting with tile

Cutting out and indenting with courses of clay tile set in lime mortar was developed as a technique by the Society for the Protection of Ancient Buildings, primarily for philosophical reasons. Repairs such as this were usually rendered or limewashed. Although rare today, these "honest repairs" are sometimes used in extensively decayed masonry.

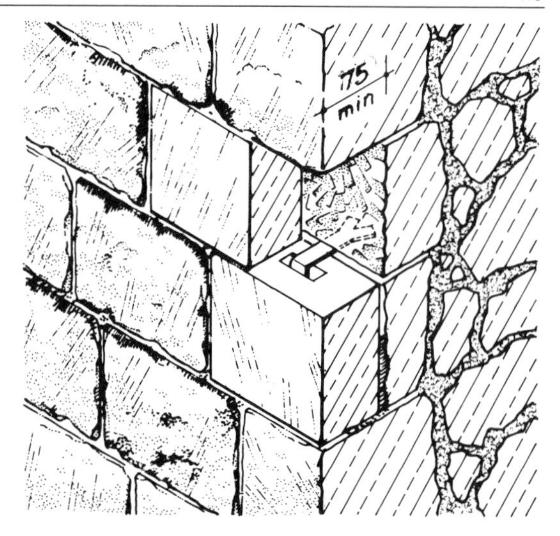

Indenting with stone

Individual stones are placed, 75-100mm on bed, into voids cut out, cleaned and sometimes bitumen-coated, on a bed of mortar. When the mortar has set, the stone is grouted through holes left in the jointing mortar. White cement, or lime and low sulphate fly ash, is the basis of the grout. Stainless steel fishtail cramps may be used as shown to tie an area of new stones back to the core.

Figure 1.1 Indenting with brick and stone

essential that the new skin should be cramped back with a staggered grid of stainless steel fishtail cramps or other suitable restraint fixing.

1.5 REPLACEMENT STONES

Matching stone

New stone should match the original as closely as possible. The help of a geologist familiar with building stones may be necessary and in many cases a substitute stone will have to be found. In these cases some knowledge of the characteristics of the original and new stones must be known. The following important references have been published by the Building Research Establishment.

- BRS Digest 268, 'The Selection of Natural Building Stones'
- D B Honeyborne, 'The Building Limestones of France', 1982
- Elaine Leary, 'The Building Limestones of the British Isles', 1983
- Elaine Leary, 'The Building Sandstones of the British Isles', 1986

Stones must be carefully matched to original sizes and profiles and, where possible, to the original finish unless the repair stone is deliberately left to a simpler profile or with a distinctive finish. Sometimes the 'original' profile may not be readily determined, especially when there has been extensive weathering or where there has been a succession of repairs and replacements perhaps over several hundred years.

To make a copy of a copy is almost always a mistake and can cause details to become less and less accurate. In such cases the advice of a competent archaeologist must be sought, so that profiles can be taken from the original stones where possible. Such information may also survive in one small, sheltered area; if so, its value is extremely high and the making of an accurate copy is essential. A profile may be drawn in situ directly onto a zinc or tough plastic sheet where this can be slipped into a joint carefully sawn out with a small masonry saw. If the joint does not occur in the run of desired moulding a fine saw cut may be made through the moulding iself. In exceptional circumstances it may be necessary to take a squeeze mould in clay and to produce a good cast from which the profile may be taken.

From these and from face measurements the bed moulds (plans of the stones), face moulds (elevations of the stones) and sections (profiles) will be prepared as drawings and as zinc or acrylic sheet templates. These drawings and templates must be carefully and indelibly marked so that their identity and location are in no doubt; they should be kept safely after the work is complete as part of the building records.

Today the replacement stones will be sawn to size and may be partly machined to reduce the time which must be spent on hand working. Finished stones should be clearly marked with a job reference and location and packed in polystyrene and straw to protect it from damage during transit and handling. Limestones and marble may receive a temporary protective slurry of lime and stone dust which can easily be cleaned down on completion of fixing. Although straw is a cheap and traditional packing it should be remembered that when wet it can stain light coloured stones; synthetic packaging is increasingly used and in many ways is to be preferred but it must be effective. To spoil expensively produced stones through carelessness is an unforgivable waste of money and shows scant regard for the work which has gone into their production. On arrival at the site the new stones must be stored off the ground to prevent absorption of water and salts from the earth, with air spaces between them and heavy-duty polyethylene sheets over them to avoid saturation from rain.

Placing new stones

Positioning and grouting

The stones will be raised into position by hand, hoist, or hand winch depending on their weight and location in the building. The cavity or open bed to receive them will have been carefully cleaned out and a mortar bed will have been spread onto the wetted, old stone. The new stone must be dampened, too, to avoid the risk of dewatering the mortar. The mortar may be a 12 mm thick bed with coarse sand and grit to match the original mortar or no more than a fine buttering with masons' putty. The stone will be handled into position and eased into the correct alignment with the aid of the lubrication provided by the wet mortar. Very heavy stones may have temporary additional support in the form of lead or slate packs.

The top bed joint and the perpendicular joints may then be stopped up on the surface leaving openings for grouting. The grout should be lime with a low sulphate fly ash (PFA) or lime and HTI powder; it should not be a neat cement grout which is brittle when set, extremely hard and notorious for staining and damage from alkali salts. Mortar staining of new light-coloured limestones is a constant problem; the recommended grouts and the protective slurry left on until completion of the work should avoid the worst risks.

Isolating paint

Where a background of core or brick cannot, for some reason, be treated with an isolating paint such as sanded bitumen the new stone may itself be painted on all but its face to avoid contamination from salt-laden moisture in the old wall. Such a treatment must stop 25 mm short of the face to avoid any risk of discolouration from the paint. The condition of the wall, the reason for the decay and likely moisture movements will influence the decision on painting, but it is generally considered to be a sensible procedure.

Bedding and pointing

New stones when not to be grouted up as described above will be bedded but not pointed until the work has 'settled in'. If the stone is a cill or lintel the bedding mortar may initially be placed under bearing points only and subsequently tamped and pointed, but this procedure relates principally to new work rather than replacement. Even so, pointing of the outer 25 mm should be left until all the bedding work has settled.

Adhesives and pinning

Sometimes new stones, or new stone faces may be secured with a polyester or epoxy adhesive. A typical example of this is the halving of decayed mullions in traceried windows, where the decayed stone is cut back to the glass line and half mullions glued to the face of the surviving internal half. Excellent as modern resin adhesives may be it is always unwise to rely on the interface bond alone. The halving technique relies, therefore, on dowel pins of stainless steel or phosphor bronze or even of glass fibre. Spalls and missing parts can be built up in phosphor bronze wire and matching mortar. The use of pins and epoxy mortars has enabled valuable masonry features, shattered as the result of bombing or fire, to be saved, which otherwise would have been lost.

The drilling and injection of holes to receive resin and reinforcement requires great care and thoughtful preparation of the site. The viscosity of the resin should permit the drilling to be filled adequately under the pressure from a gun or a hypodermic syringe, whichever is appropriate. Fine fissures may be grouted with a very thin, low-viscosity resin, but the useful mobility of such materials is also a risk; it is not possible to control or to 'pull back' once injected, so that adequate precautions in the form of latex paint 'facing', modelling clay for plugging runs and swabs and solvents must be available. Latex paint is brushed on the surface

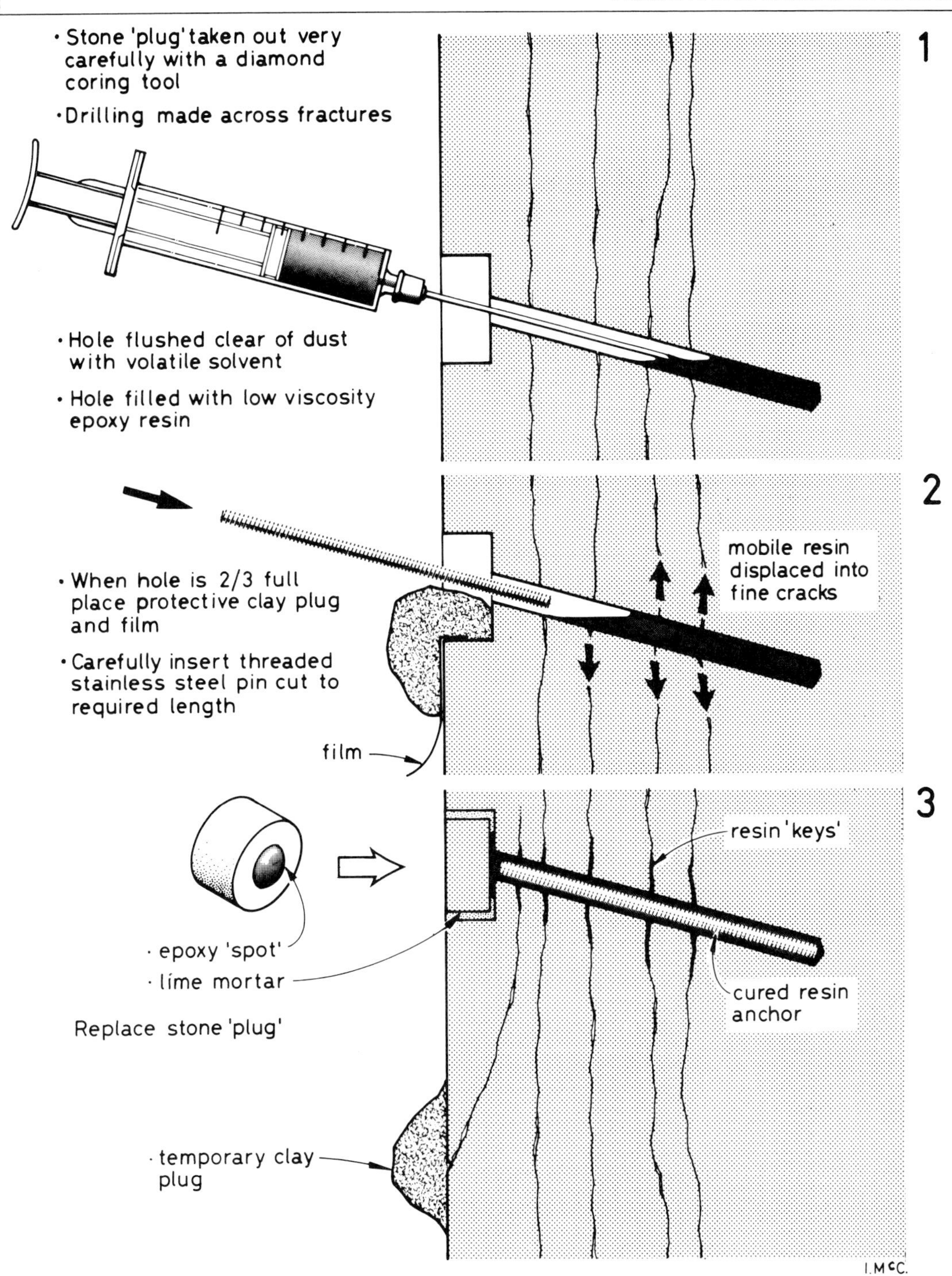

Figure 1.2 Pinning technique

in one, or preferably two, applications, which can be peeled off on completion of the work.

After holes have been drilled it is essential to remove all the dust; they must be flushed out with a solvent, or, if drying time is available, with water. Flushing out is best achieved with the same apparatus which will be used to inject the resin. Small holes may sometimes be cleared of dust by blowing out with a small tube. One of the problems associated especially with smaller holes is the entrapment of air when the resin is injected. If a hypodermic syringe is used, a length of tube or plastic drinking straw, cut to the depth of the drilling, can be attached to the end of the hypodermic and filled with resin before insertion into the hole. In this way, the hole will be filled from the deepest point back to the surface. The amount of resin injected into the hole must take account of the displacement which will occur when the reinforcement is inserted. Unless the hole is very small in diameter, the resin should not come too close to the surface. For a drilling 6 mm ($\frac{1}{4}$ in) in diameter, prepared to take a 3 mm ($\frac{1}{8}$ in) rod, the drilling should be injected for approximately $\frac{2}{3}$ of its depth. Pins should be cut to size before injecting resin: the heads of the pins should not be closer to the surface than 6 mm ($\frac{1}{4}$ in) for the small diameters, to 12 mm ($\frac{1}{2}$ in) for large diameters, allowing the outer 6–12 mm ($\frac{1}{4}$ in–$\frac{1}{2}$ in) to be filled with a fine matching mortar.

Where possible, wire pins should be turned at the end to provide a better key. Rods should be threaded for the same reason and if glass fibre, should be well roughened.

Replacement with cast stone

Although there are no savings on labour on site or on disruption for fixing when cast stone is used in place of natural stone, economies are achieved when repetitive elements need to be produced. In some situations, too, cast stone has been thought preferable to natural in replacing copings, ridges and chimney caps where the environment is particularly demanding and aggressive. Cast stone has been used extensively in England as a substitute for some forms of stone slate and paving. Casts of varying quality have also been used extensively as in situ replacements for sculpture which has been removed to some place of safety. In all cases its use should be an exception and every effort must be made to use natural stone replacement on historic work. Cast stone always weathers differently from natural stone.

1.6 REDRESSING STONE

The removal of the original face from the surface of an old stone wall is a drastic process and one which is quite alien to the normal principles of conservation. Although the practice should be resisted while there is any hope of conserving the original face, there are some circumstances when it is justified; for instance, where the face of the stones has become badly disfigured by blistering, splitting, spalling or by poor quality, superficial repairs, or where extreme deterioration is posing a threat to the general public. There are many examples where redressing has taken place on a large scale (London and Oxford), where there was no satisfactory alternative; but there are many other examples where redressing has been used as a cosmetic treatment, with the object of reintroducing uniformity and creating a new appearance. Destruction of an original face for such reasons is always to be discouraged, although there is no technical reason why it should not take place and no reason why accelerated weathering should be anticipated.

1.7 APPENDICES

Appendix 1

'The Building Limestones of the British Isles',
Building Research Establishment Report (Elaine Leary)

Statement from the Research and Technical Advisory Service, Historic Buildings and Monuments Commission for England

A number of cases of stone selection and exclusion which have been made solely on the basis of a reference to the BRE Report 'The Building Limestones of the British Isles' indicates that it is necessary to underline some of the qualifying statements of the report.

Stone selection

The obvious and general rule to be followed in replacing decayed or damaged stones is that like should be replaced with like. Lack of availability is the usual problem which introduces substitute stones and, in many cases, the need for performance data.

In the case of new buildings, the selection criteria are less obvious and, again, performance data is likely to be required, especially when the specifier is unfamiliar with building stone.

In his introduction to the report, the Director of the Building Research Establishment described as 'a bold step' the intention to indicate the comparative durability of each stone. One of the underlying reasons for this statement must relate to the fact that test results of this kind tend to be viewed most uncritically by those who are least familiar with building stone and who most need the information.

Examples of misuse of the report

Typical misuses of the report are of two kinds. The first kind involves the specification of a stone solely on the grounds of a good durability rating from the report, without any regard to environmental compatibility, historic precedent, bed heights and other natural size limitations and restricting characteristics such as vents or soft beds.

The second kind involves the exclusion of stones which, typically, fall into the 'middle zones' of durability where other factors, such as detailed observation of traditional and recent field performances tend to be at variance with test results and must be taken into consideration.

Both misuses stem from inadequate time and thought being given to selection and to a proper study of the report, which explains the problems and limitations of the testing programme.

Recommendations
When stone is required, the following steps are recommended:

1 Identify the stone to be replaced, or, in the case of a new building, what is required to suit the design elements and, where applicable, what is the traditional local use of stone
2 Look at the availability of matching stones, using the *Natural Stone Quarry Directory* (Stone Federation) and the BRE Report. Establish available sizes, workshop facilities and delivery times
3 Look at the reference buildings and recent contracts and the durability ratings in the BRE Report
4 Where the durability rating conflicts with field observations, cross-check again the reference buildings and refer to BRE or RTAS (English Heritage).

Appendix 2

French limestones for the UK

There is considerable experience now in the UK of imported building stones for use in the repair of historic buildings. Many of the these have been tested for durability (using the 14 per cent sodium sulphate crystallization test) and others are being tested from time to time.

Although the original stone should always be accurately matched where possible, the following list suggests suitable matchings for English stones which may be impossible to obtain in desired sizes and colours. All the continental matchings given are usually available in considerable bed heights.

English stone	*Imported match*
Anstrude Hard White	Anstrude Jaune Clair, Vilhonneur
Bath (Monks Park)	St Maximin Fine
Bath (Westwood type)	St Maximin Fine, Courteraie Demi-Fine
Bath (Box type)	Lavoux, St Maximin Franche Fine
Blue Lias	Anstrude Bleu
Anston	Jaumont, Anstrude Jaune
Clunch	Cleris, Richemont
Clipsham/some Lincolnshire stones	Anstrude Jaune Clair, Anstrude Jaune, Massangis Marbrier
Guiting	Jaumont
Ham Hill	Pondres, Jaumont
Oxfordshire Limestones (some)	Besace, Anstrude Jaune Clair, Savonnieres Demi-Fine

REFERENCES

1 Ashurst, J; Dimes, F G; Honeyborne, D B, *The Conservation of Building and Decorative Stone*, Butterworths Scientific, Guildford, 1988.
2 Ashurst, J and Dimes, F G, *Stone in Building: Its Use and Potential Today*, Architectural Press, London, 1977 (out of print). Reprinted by the Stone Federation, 1984.
3 British Standards Institution, *BS 5390: 1976 (1984) Code of Practice for Stone Masonry*.
4 Building Research Establishment, Digest 177, *Decay and Conservation of Stone Masonry*; Digest 269, *The Selection of Natural Building Stone*.
5 Caroe, A D R and Caroe, M B, *Stonework: Maintenance and Surface Repair*, Council for the Care of Churches, London, 1984.
6 Schaffer, R J, *The Weathering of Natural Building Stones*, Building Research Establishment, HMSO, reprinted 1972.
7 *Stone Federation Handbook*, BEC, Cavendish Street, London, 1986. (Directory of trade members, includes glossary of masonry terms.)

See also the Technical Bibliography, Volume 5.

2 CONTROL OF ORGANIC GROWTH

2.1 CONTROL OF ALGAL SLIMES, LICHENS, MOSSES

There are many circumstances in which lichen and even some varieties of small plant may enhance the appearance of masonry without adverse effect. Other circumstances require a sterilizing treatment for reasons of maintenance or appearance. Biological growths which should receive attention include unsightly algal slimes on vertical surfaces and especially on paving, lichens causing deterioration of various building materials such as above copper or lead sheet, polished marble or lime plaster surfaces and glass, and fouling organisms in drain systems.

An important point to remember in planning the consolidation and maintenance of many unroofed monuments with limited and expensive access to exposed wall tops is the function of lichens in nature as soil formers. Lichens may assist, in time, the establishment of mosses, small plants, and even trees. Complete cleansing of these areas during a consolidation programme is therefore of great importance.

Due respect should be paid to the conservation of unusual or harmless flora where control and observation are possible.

Treatments

A great variety of treatments is available which effect an initial kill. Unfortunately, some of the traditional treatments, for instance the use of weed killers incorporating calcium chloride, build up residues of damaging soluble salts. It is also unfortunate that concentrated solutions of zinc or magnesium silicofluoride may produce hard surface skins on limestone which are liable to spall off.

A long-term inhibiting effect on biological growth on some walls may be obtained by the installation of narrow flashing strips of thin-gauge copper. These are tucked into the length of horizontal joints in the masonry at approximately every 1 metre. The effect of rain washing over the strips is to subject the face of the masonry to a mildly toxic wash. A certain amount of light green stain must be anticipated which makes this system unsuitable for very light coloured stones, and of course it will not be effective where detailing on the building tends to throw off the rain.

The recommended treatment for masonry covered with algae, lichen, mosses and small plants is described below. The biocides based on a quaternary ammonium compound effect the initial kill, and when combined with tributyl tin oxide will have a long-term inhibiting effect on recolonization.

When handling and mixing biocides, remember to wear rubber gloves and in addition, when spraying, to wear safety glasses, mask and goggles. Do not spray in the immediate vicinity of other unprotected people and animals.

1. Remove as much growth as possible in the form of plants and thick cushions of moss with knife blades, spatulae and stiff bristle or non-ferrous soft wire brushes. If the surface below the growth is delicate or liable to be marked or scoured in any way this preparation must be limited to lifting of the moss only
2. Prepare a solution of quaternary ammonium-based biocide to the manufacturer's specification
3. Fill a pneumatic garden-type sprayer two-thirds full with the diluted biocide. Adjust the nozzle to a coarse spray setting. There should be sufficient pressure at the wand nozzle, after pumping the container, to saturate the surface of the masonry without causing excessive 'bounce back' and spray drift
4. Apply a flood coat. Commence at the top of the vertical surface to be treated and move across horizontally and slowly to allow approximately 100 mm run down. The next horizontal pass should be made across the previous run down
5. Leave the treated area for at least one week. Brush off as much dead growth as possible with bristle brushes, making sure that any adjacent gutters and hoppers are kept clear
6. Prepare a solution of a proprietary biocide based on a quaternary ammonium compound and incorporating tributyl tin oxide or other proven long-lasting biocide to the manufacturer's specification
7. Fill a second pneumatic sprayer with the diluted biocide and apply in the same manner as (4) above
8. Allow the surface to absorb and carry out a second application of proprietary biocide as (6) above

Protection of other areas

Provided that the applications are made carefully there should be little risk to grass or flowers below the treated area. However, as there is always a risk of spillage, it is sensible to lay a sheet over plants on the ground whilst working. In close proximity to ponds containing fish and other aquatic wild life it would be prudent to carry out mechanical cleaning alone to avoid the possibility of contamination during treatment and subsequently due to leaching of biocide from treated surfaces.

Coverage

This will vary with site conditions, but as an approximate guide, 1 litre of biocide treats 1.5 m^2. Do not prepare more biocide than is to be used in one day. There is some evidence that there will be a lessening of toxicity when diluted biocide is stored for a long time.

Failures with these treatments are not unknown, but if the above procedures are followed there should be no problems. In exceptionally dry periods it may be beneficial to revive dormant dry lichen, which tends to be water repellent, with light water spraying before applying the biocide. Application of biocidal treatments should not be undertaken during wet weather or when windy conditions lead to excessive drift of spray. It is important that products are applied in strict accordance with the manufacturer's recommendations for safety and protection of the environment.

2.2 CONTROL OF IVY AND OTHER CREEPERS

Well-established, decorative climbing plants and creepers may have a beauty and value of their own but can, unfortunately, become a real problem by limiting essential maintenance of buildings against or on which they are growing. This is, inevitably, a matter of controversy from time to time. It is always disappointing to see a flourishing, mature decorative plant cut down or destroyed and every sensible attempt should be made to preserve good plant specimens, but not at the expense of the building.

Typical problems associated with mature wall creepers or other large plants growing against walls are as follows:

- Persistently damp walls
- Disturbance of footings and plinths
- Disturbance of eaves courses
- Disturbance and blocking of rainwater disposal systems
- Scouring of soft wall surfaces
- Limiting access for painting and repair
- Restricting maintenance inspection
- Security risks (ready access to upper windows).

In the case of ivy (excepting decorative variegated varieties) the much more serious problem is its rapid growth with aerial roots intruding into joints and displacing stones or bricks. Suckers and tendrils will also contribute to surface decay, especially of mortar, by the secretion of acid substances. In the case of field monuments, where ivy grows readily and penetrates deeply into lime

Although ivy and other creepers may look attractive on masonry buildings the effect of some growths can be very destructive in maturity. In this illustration, woody root growths have penetrated lime mortar joints and with increase in their girth have jacked apart substantial limestone blocks. The control of such growths is essential and sometimes complete removal is the best answer.

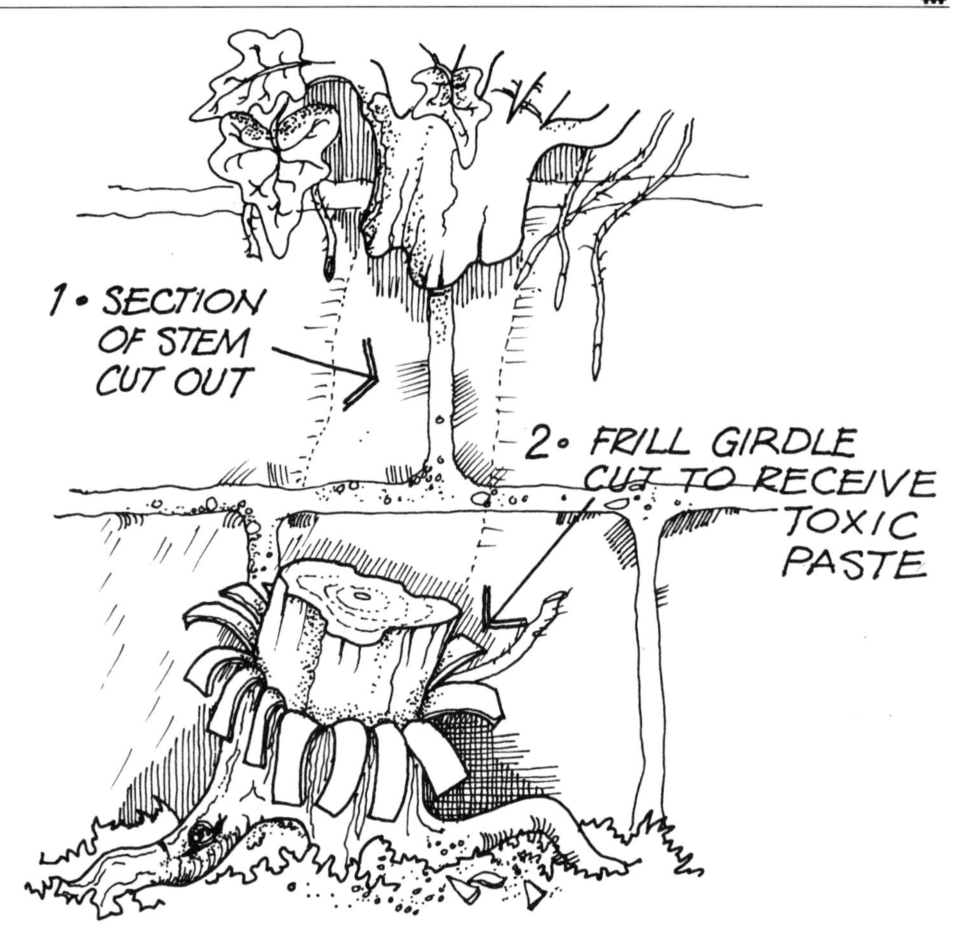

Figure 2.1 Treatment for removal of major ivy growth

corework, the stability of an entire wall can be threatened. In general, therefore, it is strongly recommended that large plants are kept away from walls unless specifically built into the landscape for the purpose. If a compromise is to be made the plant should be allowed to climb against a stainless steel angle frame strung with stainless steel wire and supported on, but separated from, the wall with brackets or long bolts in tubular spacers; growth should be well controlled and kept away from eaves, gutters and openings.

Removal of ivy

A length of the main stem of the ivy should be cut out at a convenient height above ground level. The plant may be left in this state to die of its own accord, but it is quite possible for a well-established plant to survive on the wall for up to two years after such an operation. Spraying the plant with toxic materials will hasten its destruction. Such materials are listed in Figure 2.2. Manufacturer's instructions must be carefully followed.

The parent stem, after cutting, should be cut to a frill girdle and the exposed surfaces coated with a paste made from ammonium sulphamate crystals (Figure 2.1). In this condition the root system may be left to absorb and die. This is a preferable system to drilling and pouring in a corrosive acid, but if an acid is used the drillings must be plugged afterwards. The ammonium sulphamate must not be used on masonry surfaces where, in association with lime, it would become a nitrogenous fertilizer. Once again it is important to adhere to the manufacturer's recommendations on safety and environmental protection.

The removal of the dead plant from the wall may be straightforward, but the temptation to pull off a well-established mat of vegetation with a rope must be resisted. Mature growth must be carefully cut or pulled out of every joint, and

Table 2.1
Chemicals used for the control of weeds

Active ingredients	*Proprietary name*	*Manufacturer/Supplier*
MCPA	Agroxone	ICI Ltd, Plant Protection Division
	Agritox	May and Baker Ltd
	Cornox M	The Boots Co Ltd
2, 4-D + dicamba	Evergreen Lawn Weedkiller	Fisons Horticulture Division
2, 4-D + dicamba + ioxynil	Bio Lawn Weedkiller	Pan Britannica Industries Ltd
2, 4-D + dichlorprop	Murphy Lawn	Murphy Chemical Ltd
2, 4-D fenoprop	4-50 Selective Weedkiller	Synchemicals Ltd
2, 4-D + mecoprop	Boots Lawn Weedkiller	The Boots Co Ltd
	Supertox	May and Baker Ltd
	Verdone	ICI Garden Products
Glyphosate	Murphy's Tumbleweed	Murphy Chemical Ltd

(This table contributed by BRE Princes Risborough Laboratories, Dr A F Bravery and C Grant).

wedging of the blocks or deep tamping carried out as work proceeds. Tamping, grouting and pointing and resetting of stones, especially on wall tops, must all be anticipated as remedial work. Large sections of dead wood must not be left within the core. As they decay they will remove support and create voids and weaknesses in the wall.

2.3 CONTROL OF WEEDS IN ESTABLISHED GRASSLAND*

For the selective control of broad-leaved annuals in turf and grass the phenoxy acid herbicides such as 2, 4-D and MCPA are recommended. Where a variety of weeds is present a product containing a mixture of 2, 4-D and either mecroprop, dichlorprop, fenoprop or decamba is suggested.

Perennial herbaceous and woody species of weeds are best controlled by spot spraying with glyphosate. As this is a non-selective herbicide care should be taken to protect non-target species from drift; contamination of general purpose spray equipment, etc. should also be avoided.

Details of use rates, modes and timing of application etc are available from suppliers.

*These notes have been based on recommendations obtained from the former Weeds Research Organisation, Yarnton, by BRE, Princes Risborough Laboratories.

3 GROUTING MASONRY WALLS

3.1 THE NEED FOR GROUTING

The consolidation of historic masonry often involves the need to stabilize walls by filling voids within their thickness. This operation is most commonly needed when thick walls of double skin construction with rubble core filling have been subjected to the percolation of water for many years. The tendency of this washing action is to cause the mortar (often of poor quality infills) to disintegrate and either to wash out of open joints or to accumulate as loose fill at the base of the wall or pier, sometimes causing bulging, cracking and displacement of stones. The absence of such evidence on the face, however, should not be taken to indicate a solid and stable condition within. Disintegrated joints must always be raked out and probed for voids, and 'sounding' with a hammer carried out to test for hollows. The removal of selected face stones and the drilling of deep cores of, say 100 mm diameter are other ways of investigating the core. Shallow stones placed in the wall as putlog hole fillers can often be conveniently removed to allow for exploratory coring.

3.2 DETAILED INVESTIGATIONS

In special circumstances, especially when large areas are involved, it may be considered advisable to commission a detailed specialist investigation of at least some typical areas where voids or internal fractures are suspected. Examination may be by gamma radiography (X-ray) or by ultrasonic measurements. Both methods of investigation require scaffolding for measurements above ground level. X-rays are effective for up to about 450 mm (18 in) through masonry, but this range can be extended by drilling pilot holes or by removing some of the face stones. During the working period the site must be properly roped off with warning signs displayed and the local authority informed. Night working may be necessary.

Ultrasonic methods of testing for voids and moisture are more versatile than

Massive masonry is usually constructed of two skins of dressed stone with a lime mortar and rubble filling between. The condition of this filling is usually critical to the stability of the wall or pier. This illustration shows circular piers which have been subjected to considerable washing out of their core filling. Constant percolation has widened many of the joints, a condition exacerbated in sandstone because of the decay encouraged by deposition of calcium carbonate and sulphate in the pores of the stone. Much of this masonry sounds hollow when struck, confirming that liquid mortar ('grout') is required.

X-ray, but require interpretation. Once a 'datum' signal has been established through a known solid section of the wall (a door or window jamb is often useful) other measurements may be taken through the wall on, say, a 200 mm grid and a solid to void pattern established over the elevation.

3.3 GROUT MIXTURES

The variety of grouting mixtures is as great as the variety of bedding and pointing mortars, although until comparatively recent times simple grouts consisting only of ordinary Portland cement and water were in universal use. These are not recommended. Large quantities of this kind of grout can create considerable problems for ancient masonry because of formation of partially soluble materials such as calcium and sodium hydroxide during the setting reaction. These may cause dark staining, efflorescences and local surface failures due to crystallization stresses. Although such grouts have the attraction of cheapness and simplicity the safest and simplest course is to eliminate them from work on historic fabric. Simple cement grouting of large voids is not even very efficient; as well as the soluble salt problem there are difficulties associated with poor mobility, high shrinkage and final brittleness.

One of the most useful grouting materials is undoubtedly low sulphate pulverized fuel ash ('PFA' or 'fly ash') which, over the last few decades has been used increasingly with cement, or lime, or both, and other additives which provide bulk or aid mobility and suspension. Non-hydraulic lime cannot be used without a setting aid. PFA and lime combined can produce a mobile, low- to medium-strength grout which is frequently just what is required to fill voids in double-skin, rubble cored walls. Reactive PFA is a pozzolanic additive which, in addition, aids penetration. Bentonite is another relatively cheap additive which helps to keep the cement/ash/lime material in suspension, avoiding 'settling out' during the grouting process.

Unless the grouting operation is small or very specialized, it is recommended that pre-bagged grout mixtures are used as much as possible. These are available for a very wide range of functions and specialist suppliers will recommend grouts for particular situations.

For standard void filling in historic fabric a pre-bagged lime:PFA:bentonite mix of 1:1:$\frac{1}{2}$ is often suitable. A higher strength grout may be supplied as, for example, low sulphate cement:lime:PFA 1:2:1. Solids to water ratios are, typically 1:3 or 1:4. More complex, more sophisticated grouts should be discussed with the supplier.

3.4 GROUTING OPERATIONS

The use of liquid grout avoids dismantling and rebuilding defective masonry in many cases. In its simplest form grouting may be carried out by hand pouring into clay grout cups formed on the face of a wall. There is a choice of four basic methods which will be dictated by the nature and condition of the masonry.

Gravity grouting is particularly suitable where the masonry is very vulnerable to movement under pressure and is the system most commonly used on ancient monument work. Pumped systems of various kinds may be used to deal with most grouting problems. Vacuum systems may be useful where fine fractures and small-scale voids are suspected.

Hand grouting

Local grouting can be carried out very efficiently by hand. This technique is suited to small, isolated voids or fine cracks and is frequently carried out in association with tamping and pointing.

The traditional hand-grouting technique makes use of small clay grout cups ('swallows' nests') formed in modelling clay against the masonry surface (Figure 3.1). Grout is poured in and allowed to disperse through the void around which the cup is formed. The grout is topped up, normally, until the level is held. When the grout has set, the cup and residue of grout may be broken off the wall and the surface brushed down. Cutting out and pointing follow.

Grouts used in this manner are commonly of the same type used in gravity systems, including hydraulic lime:sand mixtures. Flushing out with water must, of course, precede the grouting in the usual manner.

Smaller cavities and fractures may be filled using hypodermic syringes. Finely ground hydraulic lime and PFA or brick dust may be used through large syringes. Patterns of small voids, such as cracks along the top of bed joints, may be filled by drilling and inserting self-sealing grouting dowels, fitted with nipples. A hand-grouting gun may be filled with a grout such as hydraulic lime and brick dust gauged with an acrylic emulsion as the vehicle. Grouts of this kind are highly mobile and may be forced considerable distances by hand.

Cracks may also be grouted through syringes or through guns and grout dowels with polyster and epoxy resins, especially where sealing and re-adhesion are required. The viscosity and strength of such materials can be modified for particular requirements. Their cost is high and may be prohibitively so on large areas.

Gravity grouting

The grouting apparatus required for filling large voids consists of one or two open galvanized iron pans with outlets in the bottoms. A union with a 38 mm diameter galvanized pipe is fitted to the outlet, which in turn is connected by means of couplings to several lengths of 38 mm diameter hose terminating in a galvanized iron nozzle 18 mm in diameter fitted with a stopcock; each grout pan is provided with a wooden plug about 450 mm long to fit into the hole in the pan bottom, and with a plunger in the form of a rubber cup on a wooden handle. This plunger is used when the grout is flowing to give an added impetus to the flow in the event of an airlock or other stoppage in the tube.

Preparation

Small holes are drilled into the wall where voids have been located or are anticipated. They should be about one metre (36 in) apart horizontally and

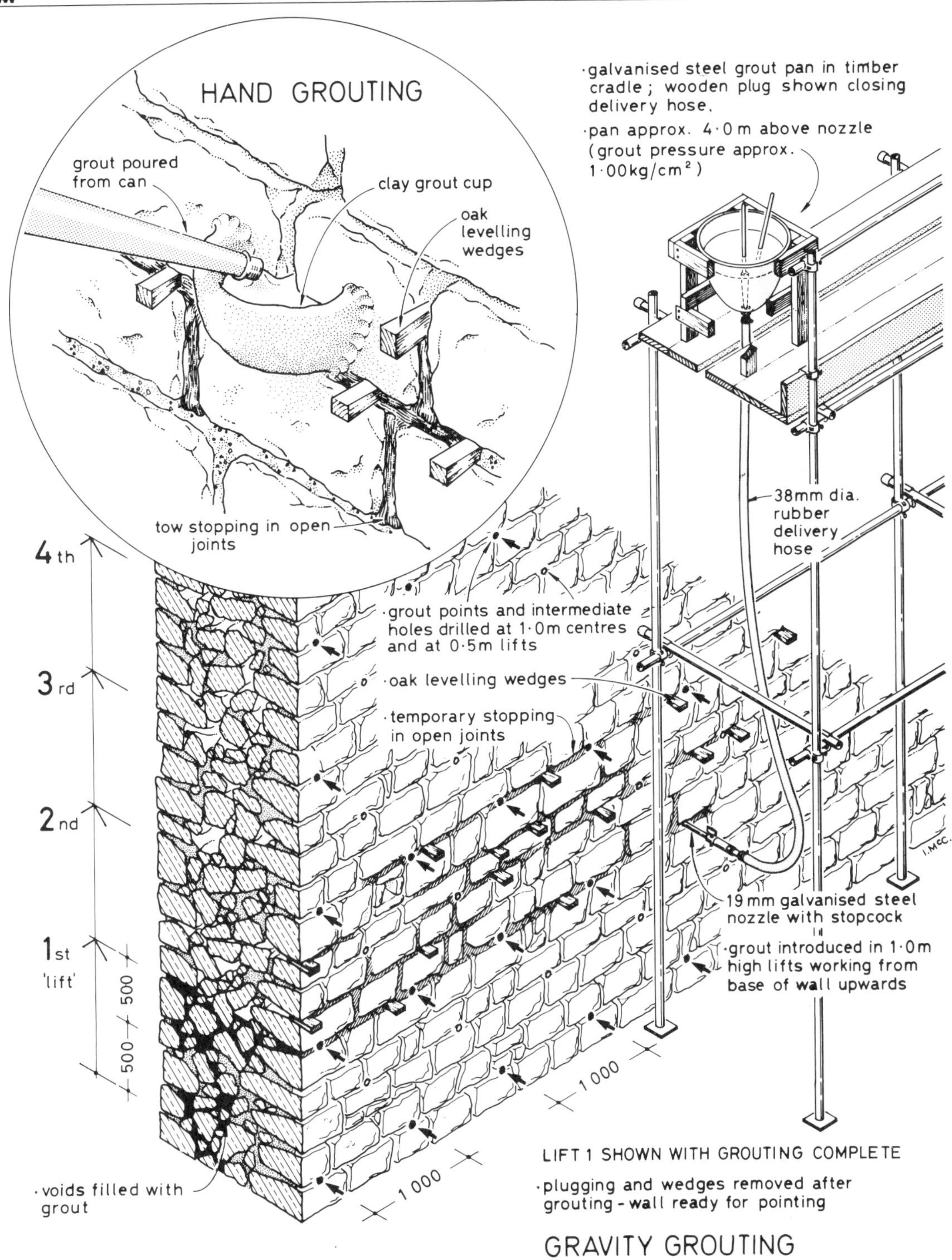

Figure 3.1 Hand and gravity grouting

450 mm (18 in) vertically on a staggered pattern. As the holes are drilled they should be washed out thoroughly with clean water, pouring in at the top holes and continuing to pour until the water runs out clean at the bottom. During this process a note should be taken of the joints through which the water runs out; before grouting is commenced these joints must be tightly filled with tow or clay, pressed well into a depth of 40 mm–50 mm (1½ in–2 in). The nozzle of the delivery hose is then inserted into the lowest hole and plugged round with tow.

Operation

To operate the simple equipment two men are stationed at the upper level with the grout pans regulating the flow of grout into the delivery hose from one pan and mixing the grout in the second pan ready for use, so that a continuous operation can be carried out. A third man is stationed at the lower level to open and close the stopcock on the nozzle as required. Ample supplies of water and grout components must be kept on the scaffold. When the grout has been mixed to the right flow consistency in the pan the wooden plug is withdrawn and the grout flows down the delivery hose. The stopcock on the nozzle is opened allowing the grout to flow into the wall until the grout level in the wall has risen sufficiently to begin to flow out of the series of holes immediately above. These holes may then be stopped up, the grout cut off, and another section of wall prepared or grouted while the first begins to set. After the initial set the tow or clay can be stripped out of the joints in readiness for pointing at a later stage. The next lift can then be grouted in the same way. One metre lift (36 in) should be taken as the maximum lift at a time to avoid the build up of pressure from liquid grout behind loose face stones. A pressure of about 70–80 kPa (10–15 psi) is obtained in the hose with the pan placed about 3.5 m–4.5 m (11½ ft–15 ft) above the point of inlet.

An accurate record must always be kept.

Pumped systems

Hand- and power-operated pumps usually consist of a mixer diaphragm pump, suction and delivery hoses and metal nozzles fitted with stopcocks. Hand-operated pumps are recommended for ancient masonry in unstable conditions. The compact nature of these assemblies usually permits the plant to be located adjacent to the work in progress and cuts down on the hose lengths required. required.

Preparation

This is generally as described under the gravity system. The nozzles are fitted into the holes and plugged around with tow. The lowest nozzle (usually) is then coupled up to the delivery hose.

Operation

One man will be required to operate the mixer, one to operate the pump, and one to open and close the stopcock as required. When all is ready the stopcock is opened on the nozzle and the pump started. The level of the grout rising up the wall is indicated by the seepage of grout from weep holes which can then be plugged with clay. Hidden grout flows may sometimes be identified by the wall surface 'sweating', as water is forced through under pressure.

When the grout reaches the next line of nozzles, the lower stopcock can be closed, the delivery hose removed and coupled to the nozzle above. The lower nozzle can be left in position until the grout has set.

The maximum pressure obtained depends upon the model being used, but a range of 70–280 kPa (10–40 psi) is usual. Much lower pressures are obtained with hand-operated pumps. Hand-operated pumps have a capacity of 18–45 litres/minute (4–10 gallons/minute). Power-operated pumps have a capacity of in the region of 1300–1800 litres/hour (290–400 gallons/hour).

Aerated pressure system

This system is useful in large-scale grouting especially where tunnels and vaults are involved. The apparatus consists of a compressor, mixer, pressure vessel, air lines and delivery hose with a wide variety of nozzle designs suitable both for pointing and grouting. The pointing finish is unsatisfactory and messy if left from the nozzle, but can be acceptable if followed up with pointing tools.

The preparation of the walls for grouting is the same as that used in the gravity system. Metal nozzles are fixed into drilled holes and plugged round with tow. The spacing of the holes will vary with the condition of the masonry, but could be set, for example, 450 mm (18 in) apart and 1.25 m (45 in) apart horizontally. The point positions should be staggered as before.

During operation, one man is stationed at the nozzles to open and close the stopcocks, one man at the pressure vessel to ensure that the correct pressure is maintained, and one man at the mixer preparing the next grout batch.

Vacuum systems

These entail enclosing the area to be grouted in an airtight transparent, flexible shroud such as a polyethylene sheet. Air is evacuated from the shrouded area by means of an air line to a powerful vacuum pump and a tap opened to allow the material from the grout pan to be sucked in. When impregnation is complete as observed through the polyethylene the tap is closed until the grouting material is set.

In spite of the obvious attractions of grouting under vacuum the practical difficulties on site must be acknowledged and not underestimated. The application to masonry structures is, at the present time, very limited.

4 REPAIR WITH MORTAR ('PLASTIC REPAIR')

4.1 THE DEVELOPMENT OF PLASTIC REPAIR

Natural stones, and to a lesser extent bricks and tiles, have been copied in wet mixes of binder and aggregates since deterioration and decay first became disfigurement and maintenance problems. Hence there are examples of ancient classical and medieval mortar repairs as well as the more familiar nineteenth- and twentieth-century ones. Of course, there have been reasons other than economy for selecting mortar or plaster as the repair medium: non-availability of the original material and fashion have both produced artificial bricks and stones of one kind or another.

Some of the historical substitutes for stone are, by now, themselves the subject of conservation. The last century saw the development of a host of artificial stones, some patented. These were usually, in the last half of the century, compositions of Portland cement and fine aggregates, including stone or brick aggregates of the type being matched, in typical proportions of 2:7. Pigments were frequently included and cast blocks were carved while still green. Paraffin oil and French chalk were commonly used as release agents. A few of these artificial stones were, no doubt, good matches, but most were too strong and subject to crazing and loss of adhesion; rarely have they weathered, except in the driest and kindest of environments, to any real semblance of the material they sought to copy.

The most common form of imitation stone or brick is the coloured binder and aggregate mixture known popularly as 'plastic stone'.

4.2 DECIDING TO USE PLASTIC REPAIR

Repair with mortars, or 'plastic' repair as it is traditionally known, is a useful technique which may sometimes be used as an alternative to cutting out and

'Plastic' stone has a bad reputation because much of it has been carried out in an amateur way and on too large a scale. This illustration shows what should NOT happen. A dense, impermeable mortar, coloured with pigment and exhibiting trowel and float marks has been plastered onto decaying stone quoins. It is relying on a bonding agent for adhesion. Around the feather-edges of this, mortar decay of the original stone is already accelerating.

piecing in with new stone or brick. Unfortunately the reputation of such repairs has suffered from inadequate specification, misuse and inexpert handling. Plastic repair is thought of as a cheap option to repairing with the natural material but the cheapness relates very often to poor-quality workmanship. Properly prepared and placed plastic repair is not cheap, except that its use may sometimes mean the avoidance of expensive items such as the temporary supports for vault and arch stone replacements and extensive cutting out.

Plastic repairs are of particular interest and importance to conservators because the technique frequently permits the retention of more original material with much less disturbance than would be possible for the execution of conventional masonry repairs. In this respect the familiar description of the method as 'dentistry repair' is very apt, involving the careful removal of decayed material, the cleaning and sterilization of the cavity and the placing, compaction and finishing of amalgam.

The analogy with dentistry may be extended further, because if careless filling of imperfectly prepared cavities is carried out much energy and expenditure will have been wasted and failure will occur in a predictably short time. Failure of plastic repairs may be both cosmetic and mechanical. In particular, mortar repair material coloured with pigments, or feather-edged to ragged areas of decay, or repairs finished with steel trowels are often visually disastrous; over-strong mortars, mortars relying on bonding agents instead of mechanical keying and large surface areas in exposed positions will cause mechanical failures.

Although there are exceptions, plastic repair should always be carried out by a stone mason or a stone conservator, because their familiarity with the material should give them a feeling for the repair which other trades and disciplines will not necessarily have.

Criteria to be considered

The following criteria should affect the decision to use a plastic repair:

- Will the use of mortar enable more original material to be retained than if stone is used?
- Will the use of mortar avoid disturbing critically fragile areas?
- Will the use of mortar avoid the removal of structural elements such as vault or other arch voussoirs?
- Will mortar perform satisfactorily in the intended context (that is, is it capable of weathering adequately)? Would natural or even cast stone be more appropriate?
- Are the areas to be repaired small enough to be repaired with mortar or would rendering be more appropriate or should a large replacement of stone with matching stone be accepted?
- Will mortar provide a visually better repair than new stone or brick in the context of heavily weathered, softened outlines?
- Are the appropriate skills available to produce high-quality mortar repairs?

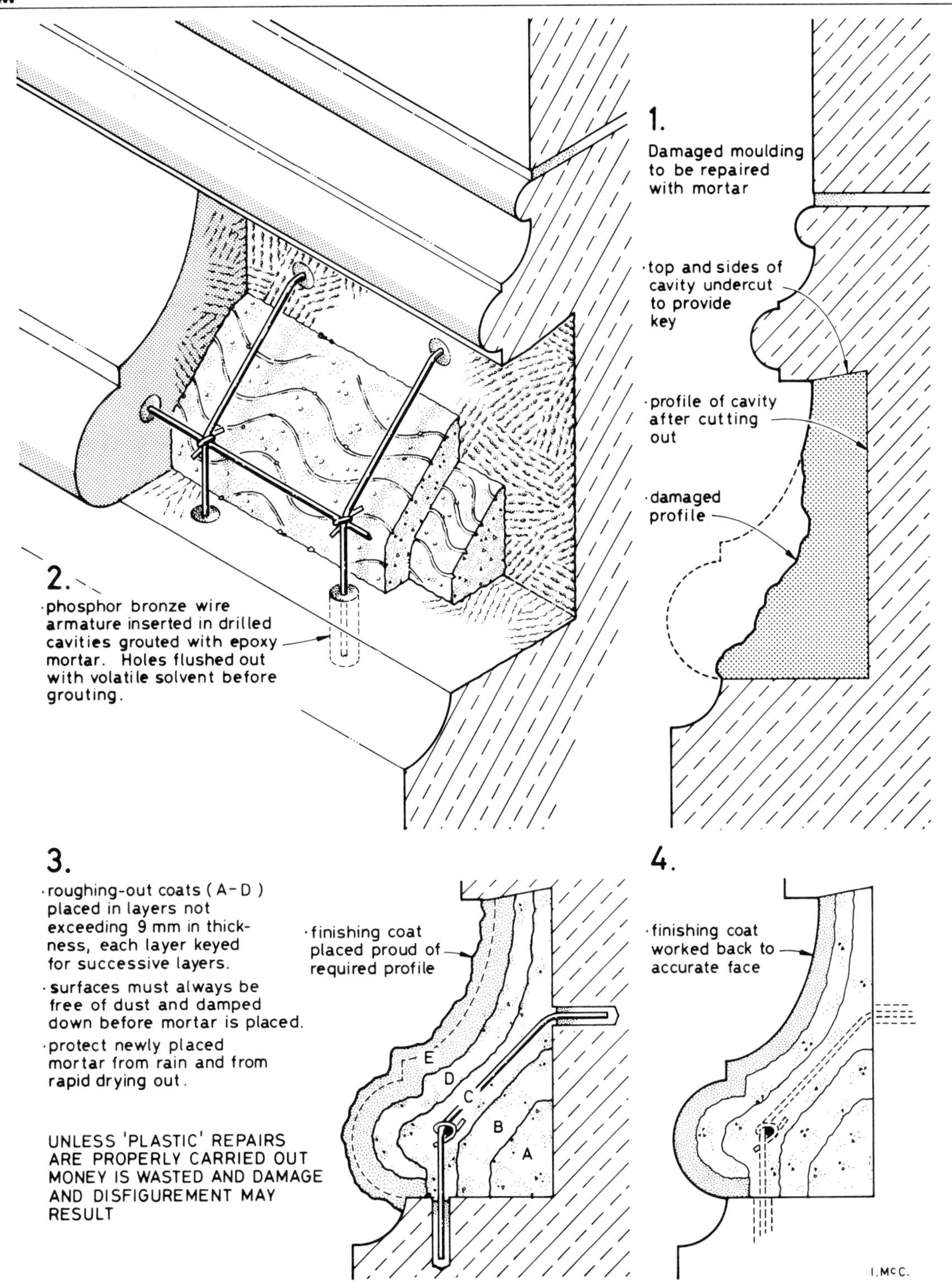

Figure 4.1 Plastic repair technique

4.3 PLASTIC REPAIR TECHNIQUE

If after consideration of the above factors, it is decided to proceed with mortar repairs, wholly or partially, the following procedures, which are illustrated at the end of this note, should be put into operation.

1. Prepare a schedule of bricks or stones to be prepared with mortar (or with brick or stone)
2. Prepare samples of mortar to match the various conditions of weathering on the building. (Weathered stones and bricks exhibit a subtle variety of colour which must be matched in the repairs. Much plastic repair suffers from an unnatural uniformity.) The repairs must be prepared as samples on a piece of stone or tile, not in a wooden mould, and judged on the dry and wet appearance
3. Cut out the decayed areas
4. Wash and sterilize the cavity with water and formalin
5. Saturate the cavity with water using hand sprays to prevent dewatering of the repair mortar
6. Place the selected repair mortar, compacting in layers not exceeding 10 mm in thickness in any one application. Allow each layer to dry out before rewetting and placing the next.
7. In cavities exceeding 50 mm in depth and extending over 50 mm^2 surface area, drill and fix non-ferrous or stainless steel reinforcement. These may range from simple pins to complex armatures. The most common materials are copper, phosphor bronze and stainless steel wire. After drilling to receive the reinforcement the holes are filled with an epoxy mortar before embedding the wire. Drilling should attempt to enlarge the cavity slightly to form a dovetail key. Holes must be flushed out with white spirit and allowed to dry before fixing with epoxy. 20 mm of cover should be allowed for any reinforcement
8. As an alternative to (7) and where the building up is of larger areas, dub out in mortar and stone spalls or cut out and build up in clay tile pieces set with mortar joints. Small structural repairs can be carried out in coursed tile which is subsequently covered with the unfinished plastic repair mortar (See Figure 1.1)
9. The repair may be finished directly to the required profile using a wood or felt-covered float, or with a damp sponge or coarse cloth. Ingenuity will provide other finishing tools appropriate to the texture of the finish required. Unsuitable tools to be avoided are steel trowels or dry, absorbent pads; the first will leave an undesirable and unnatural laitence on the surface and the second will risk the removal of water from the repair too soon. An alternative repair finishing method is to build the repair up proud of the required profile and then to work it back after an initial set has commenced on the surface with a fine saw blade or purpose-made scrapers

Additional procedures

Individual stones must be repaired separately and if necessary pointing between repairs must be carried out separately. Areas should never be repaired as a render and joint-scored.

Mortar repairs must be protected from direct sun or other rapid drying conditions. This may be achieved with damp cotton wool pads on small-scale repairs or with damp sacks on larger areas. Care taken during preparation and after placing of the repair will avoid one of the most common problems associated with this kind of work, the appearance of fine, shrinkage cracks during drying.

A 'bank' of sand and stone dusts should be set up to provide the required colours without the use of pigments. Different mixes may not always provide quite the variation in colour required and in this situation stone dusts may be added to the face of the repair before a set commences. This is very skilled work and is best excluded unless the operative is particularly experienced.

Proprietary mixes

A number of proprietary mortar repair materials are available. Some of these have shown themselves to behave well on weathering and they may be useful where on-site matching expertise is questionable. Unfortunately, to be successful, the repair mix has to be matched in a laboratory to samples of stone provided by the client and it is difficult to vary the potential strength of the repair. A match which involved a number of site visits from a laboratory would become very costly. There is no doubt that the most desirable way of forming mortar repairs is to use on-site expertise throughout.

There are also a number of proprietary repair materials available which are based on epoxy resin. While some suppliers are able to achieve good aggregate matches, the epoxy resin binder will mean the patch will be particularly noticeable when the wall is wet. To avoid this contrast, it is often suggested that the whole surface be treated with a water repellent once the repairs are completed. Such treatments are often unnecessary and should be avoided.

4.4 MIXES FOR PLASTIC REPAIRS

Lime and cement binders

Most of the plastic repair mortars are based on a lime binder, but repairs to sandstones and bricks may be better carried out using a cement binder and a plasticizer, or a masonry cement. This is because sandstone which is already decaying may further deteriorate in the presence of lime washing into the edges of the prepared cavities and because the strong colour of some sandstones and bricks are not easily matched when lime is included.

A high proportion of aggregate to binder in the cement:aggregate mixes is a further advantage when matching a strong-coloured sandstone. The mortar preparation must always ensure that the grains of sand and stonedust are adequately coated with the binder paste. When lime is used it should be prepared as putty either retained from the slaking process or run to putty by soaking white, non-hydraulic hydrated lime in water for twenty-four hours. The putty should be mixed wet with the required aggregates as far in advance of the work as possible. (See also Volume 3, Chapter 1 'Non-hydraulic lime'.) A convenient way of storing this wet mixture of lime and aggregate 'coarse stuff' is in a plastic dustbin under wet felt and an airtight lid. Although these mixtures may emerge as damp, rather crumbly material after weeks or months of storage they may be made soft, wet and plastic again by beating and ramming without the addition of more water. This is exactly what is required for a good, low-shrinkage mortar. Coarse stuff of this kind may be quite strong enough for small, sheltered repairs and even for more ambitious ones exceeding 10 mm at a time.

If the exposure is demanding, a higher-strength mortar will be required using hydraulic additives. These are listed below and should be added to and thoroughly mixed with the coarse stuff immediately before use. Hydraulic materials cannot, of course, be stored wet.

Hydraulic additives in order of increasing strength:

Finely powdered brick dust

'HTI' powder, prepared from refactory bricks

Hydraulic lime

Some PFAs

White cement

Masonry cement

Ordinary Portland cement.

'Strength' of a mortar in itself is unimportant; resistance to wetting–drying cycles may be very important indeed. A compromise must be made between the mortar strength closest to that of the surviving stone or brick and the strength considered necessary for a particular exposure. Similar properties of porosity and water absorption are also important and test cubes may be informative on this point as an ad hoc test, observing the amounts of water taken up into cubes of mortar and stone set in a test tray.

In no circumstances should a repair mix be selected on exposure grounds alone where its strength may adversely affect the condition of the adjacent original fabric. The repair should be designed to fail in advance of the material it is repairing.

Epoxy resin binders

Plastic repair mixes for sandstone based on aggregate and very small amounts of epoxy resin binder can be successful although there are very few situations in the UK where this has been achieved.

Epoxy resin repairs based on 12–18 per cent epoxy resin, by weight, began to be used in Australia ten years ago and continue to perform well. This proportion of resin is able to recreate the natural stone structure by 'gluing' the aggregate

Examples of repair mixes

The following mixes are examples of mortar repairs intended *for guidance only*. All measures are by volume. Grading of the aggregates is as critical to the successful performance of plastic repair as the binder:aggregate ratio.

Brick repair mix

Masonry cement: sharp sand: soft staining sand

1 : 3½ : 1½

(parts by volume)

Sandstone repair mix

Hydraulic lime: sharp sand: soft staining sand

1 : 2 : 1

(parts by volume)

Limestone repair mix

White cement: lime putty: aggregate

1 : 2 : 10

(parts by volume)

Note method of preparation: In this case lime and aggregates are prepared first as wet coarse stuff as follows:

1 lime putty: 2½ stone dust: 2½ sharp sand

To 10 parts of this coarse stuff, 1 part of cement is added to give the required 1:2:10 cement:lime:aggregate mix.

Good grading of aggregates – limestone repair

Lime	(1.18 mesh)	2 parts	*binder*	2 parts
Stonedust	(1.18 mesh)	1½ parts	"	
Sharp sand	(600 mm mesh)	2½ parts	*aggregates*	5 parts
Red soft sand	(400 mm mesh)	¾ part	"	
Yellow soft sand	(300 mm mesh)	¾ part	"	
Yellow brickdust	(600 mm mesh)	½ part	*pozzolanic*	
Yellow brickdust	(300 mm mesh)	1 part	*additives*	1½ parts

Total mix proportion: 2 : 6½ binder:aggregate

together at points of contact and leaving the pores unfilled. These repairs seem to have similar properties of porosity and thermal expansion to the surrounding natural stone. It is hoped that research initiated in the UK by the Research and

Technical Advisory Service of English Heritage will be able to quantify such matters in the near future.

As for other types of plastic repair mix, the matters of preparation and application technique are of paramount importance.

4.5 PROTECTION OF PLASTIC REPAIRS

Although plastic repairs should not be used for areas of extreme exposure it may be possible to use them in strings and cornices if a lead flashing is provided as well. A limited amount of experience is indicating that plastic repairs of exposed elements such as balustraded parapets may perform well if subsequently treated with a catalysed silane or other consolidant as described in Chapter 9.

5 MASONRY CLEANING

5.1 THE CASE FOR CLEANING

Building cleaning is most frequently undertaken for aesthetic reasons, although there are sometimes sound practical grounds for removing dirt when, for instance, decay is taking place around encrustations and cracks or open joints are being obscured. Buildings may also be cleaned solely or principally because repairs have become essential and access in the form of expensive scaffolding is available.

Whatever the final decision is on the desirability of cleaning, some basic questions need to be answered:

1 *Is the soiled appearance of the building worse than the cleaned appearance will be?*
To answer this question it is necessary to look carefully at the various building elements, know the identity and to understand the weathering characteristics of the masonry materials and how their colour may have changed, to consider the joints and repairs which may be largely concealed by dirt. Sometimes a fairly uniformly soiled building may be revealed as an untidy patchwork after cleaning. This does not mean, necessarily, that cleaning should not take place, but that cleaning alone will be inadequate.

2 *What kind of dirt is adhering to the building and is it causing or contributing to the deterioration of the host surfaces?*
Heavy encrustations may be seen as contributing to decay on almost all masonry surfaces. Calcium carbonate and calcium sulphate deposits on sandstone and brickwork are damaging. Street-level soiling may be seen as contributing to staining of the surfaces on which it forms. Other types and degrees of soiling may represent no particular problem.

3 *What methods of cleaning are suitable for removing the dirt deposits from the particular surfaces in question and is the requisite skill and experience available to execute the work correctly without damage?*

The first part of this question may be more easily answered than the second, by consulting these notes or other advisory material or from personal experience; but both parts may need trial cleaning samples to be carried out.

5.2 CLEANING TRIALS

A cleaning trial is not a free demonstration by a contractor who is, understandably, hoping to secure the main cleaning contract. The sample clean is an important part of a feasibility study and if it is to produce information of value without obligation it must be properly specified, recorded, supervised and paid for.

The nature of the cleaning trial is, obviously, exploratory, so that the specification will cover general requirements and limitations. The trial should assist in the following ways:

- Whether or not cleaning is aesthetically desirable and practically and economically feasible
- Definition of successful and elimination of unsatisfactory processes
- Production of detailed specification covering building preparation and protection, materials, equipment and techniques, operatives, surface repair and treatment requirements, timing and costs.

The achievement of these objectives requires the selection of a representative area or areas of the building which include as many typical conditions as possible, including openings and enrichments. Good photographic recording before, during and after operations and frequent supervision and dialogue between the specifier and the operatives are also essential. All data relating to chemical types and strengths, contact times, nozzle sizes, air pressures, abrasives and water flow rates must be recorded.

A contractor or contractors with good experience of all cleaning systems needs to be selected for the trial cleaning.

5.3 CLEANING SYSTEMS

Methods of masonry cleaning may be summarized as follows:

- *Washing*: principally limestone, marble, polished granite and dense brick
- *Mechanical*: principally sandstones, but often as a supplementary method on limestone and marble
- *Chemical*: principally sandstones, brickwork, terracotta but sometimes as a supplementary method on limestone and marble
- *Special cleaning techniques such as poulticing*: all materials.

Reasons for cleaning should always be established and the likely benefits and risks assessed. On the Bath limestone terrace shown in this illustration simple water washing has enormously enhanced the appearance of the building on the right and has improved its condition by removing heavy deposits of soot and encrustations developing in sheltered zones on blistering sulphate skins. Sometimes, however, visual benefits may only be marginal and the cleaning process may involve too many risks to justify it.

These methods are now discussed in detail. See also page 58 'Appropriate cleaning methods' for a further summary.

5.4 WASHING

Unlike sandstone, the dirt which forms on limestone and marble is soluble in water. Washing, whether by bucket and brush, multiple sprays, water lances or wet packs, is, therefore, a well established method of cleaning these surfaces.

Washing is a very simple process in principle, requiring only some sensible means of putting enough water in contact with the dirt deposits to wash them away directly, or to soften them sufficiently to allow their release by brushing. Most problems associated with washing have to do with saturation. Unfortunately, most spray cleaning systems are insufficiently versatile to deal effectively with the different degrees of soiling usually found on a building with openings, projections and enrichments. While a flat surface may be cleaned in two hours, two or three days of spraying may be required for a carved cornice with encrusted dirt. After a period of washing of this length, there may be a number of undesirable consequences.

Problems of over saturation

Unless there is provision to modify the washing programme to avoid unnecessary general saturation, the consequences may be as follows:

1 Light to dark brown staining will take place as dirty water dries out from the stones and joints, especially on new stones pieced in. In these situations it is better to clean *before* repairing wherever possible

2 Staining, usually brown, and white efflorescences may appear as a result of salt migration to the surface

3 The release of small flakes of stone, especially on small-scale and undercut detail, may occur as a result of the dissolution of salts. Such flakes may sometimes be attached to the surface only by water soluble material present as the result of the activity of a polluted environment on the stone surfaces. Surface losses will also be incurred when the stone is powdery

4 Weak jointing material may be washed out

5 Water penetration through defective joints, or through cracks and contact with iron fixings, plaster, beam ends, bond timbers, panelling, electrical fittings and furnishings may take place. There is also the risk of filling unsuspected reservoirs above vaults, or in floor spaces or in basements which may result in direct damage or future problems with dry rot (serpula lacrymans)

6 The development of disfiguring green, red or orange algae on recently washed surfaces, especially flat or inclined catchment areas may be noticed

7 In wintry conditions, considerable damage can result from the freezing of water trapped in the joints or in the pores of the stones. Ideally, no washing should take place during months likely to be frosty. If, in emergencies, such washing must continue, the work must be halted before sunset and be fully covered up. In exceptional circumstances, background heating on the scaffold will be necessary

Washing with minimum risk

Successful washing programmes are clearly those with sufficient versatility to put the minimum amount of water for the minimum time exactly where it is required and nowhere else. This may be achieved in different ways.

Traditionally, *water jets* tied to the scaffold have been turned on and shut down as the surface response dictates, but this is difficult to control in practice and groups of sprays tend to be left on for as long as dirt remains in the most stubborn areas. An experienced cleaner will get to work with small brushes of bristle, phosphor bronze or brass wire as soon as possible, to cut down the saturation period. Steel wire brushes should never be used, because of their harsh action, usually inappropriate design and the risk of leaving steel fragments on the building surface, which will later produce small, but vivid rust stains.

The ideal condition for washing is a *persistent wet mist* over the soiled face of the building, avoiding the impact effect which large water droplets have when delivered by coarse sprays. To achieve the mist, or 'fogging', the sprays need to be atomized from fine nozzles situated at least 300 mm away from the masonry face. Enough water pressure and small enough orifices are required to atomize the water. In practice, this is rarely easy, as even on a tightly sheeted scaffold, draughts of air can carry the water mist away from the building and the effectiveness of the system therefore depends on how successfully the mist can be contained.

'Intermittent' or 'pulse' washing is a more dependable system of washing which makes use of atomized sprays of water playing intermittently on the building surface. The water is controlled electronically by means of sensor heads or a preset clock. The sensor heads comprise twin carbon rods set in a non-conductive plastic body linked to an electrical control box. The sensors, about 20 mm ($\frac{3}{4}$ in) long, are pinned at intervals into the masonry joints with stainless steel staples. When a water bridge forms on the sensor head, the water is automatically cut off; when the water bridge is broken by drying out, the sprays will automatically be switched on again. In most cases, however, the clock control is preferred because it is more positive. In this case, a washing interval will be established by preliminary trial and error. The clock should be set to create spraying times of say, eight seconds, interspersed by four-minute shut-downs. The aim with either system is to supply just enough water to progressively soften the dirt without causing saturation and risking penetration through vulnerable areas. Close attendance is still required to commence scrubbing as soon as the dirt becomes responsive to brushing.

A more recent development in washing is the use of *flexible bars to position the nozzles* and direct the water exactly where it is required. Plastic-sleeved flexible bars are fixed to the scaffold with brackets and swivel mountings. These bars provide the support for the nozzles, with the result that a true three-dimensional flexibility in positioning water sprays is achieved, enabling the underside of a moulding or soffit of a niche to be cleaned, selectively, as easily as a straight run of ashlar.

The variety in design of *brushes* has increased. Several sizes are needed for all but the simplest facade, of both scrubbing and stencil type. A tight formation of phosphor bronze crinkle wire has shown itself to be particularly useful. Small blocks of sandstone, 'rubbing' or abrasive stones, are also used to remove stubborn staining and encrustation from flat surfaces.

Water penetration risks may further be reduced, especially on high buildings, by the construction of *'splash' or slurry boards* of resin bonded plywood, sheathed in polyethylene sheet, at intervals to form horizontal catchments to collect and carry off the water flowing down the building face. The water from these boards is collected in plastic gutters fixed at their outer edge and carried off in plastic downpipes. Similar constructions may be formed over large openings which are to remain in use during the cleaning.

Cold water direct from the mains is normally used for cleaning facades of buildings. For particular situations, hot water may be justified, especially where detergents or 'de-greasing' chemicals are used.

Water lances

Light soiling, especially where a high proportion is organic, may sometimes be removed by water lances without any preliminary softening up by water spraying. Alternatively they may be used in combination with water sprays, mechanical or chemical cleaning. Pressures are likely to be in the 100–115 kPa (700–800 psi) range, and the lance will usually be specified as a 'low volume, high pressure' lance. Water at these pressures has a cutting action and the design of the outlet and the technique of the operative both significantly affect the economy and the safety of the clean. Potential damage from indiscreet or careless lancing is nevertheless a factor to be taken into consideration, especially on stones with soft, sandy or clay and sand beds.

Fine sand and water can be mixed together at source for delivery through a lance at comparatively low pressures of between 2.5–4.0 kPa (18–30 psi). The water, carrying the abrasive, has a light scouring action; the abrasive can then be cut out to permit flushing with water alone. In common with conventional wet sand blasting, it will also clean sandstone.

Steam cleaning

Steam was used quite extensively before the last war when it fell into disrepute partly because caustic soda, added to the boiler water to avoid furring, was deposited on the cleaned surface and remained there to cause decay. Because steam condenses so quickly the method is little more than a hot water wash with rapid drying. Hot water is no more effective than cold in removing atmospheric dirt and so there is little point in its use. However, a combination of steam and high pressure has proved successful in removing trodden-in chewing gum from paving, and hot water in conjunction with a neutral pH soap is useful where greasy dirt is present.

5.5 MECHANICAL CLEANING

Brushing, scraping, spinning-off and 'polishing'

Mechanical cleaning removes dirt by abrading the surface. The simplest form is *dry brushing*, a technique which will remove loosely bound dirt and organic growth but little else. A harsher method is *scraping* the surface down with a suitable tool such as a mason's drag, sometimes used to remove paint from flat surfaces. The most damaging cleaning in this category in the past, however, has undoubtedly been '*spinning-off*'. This normally entails the use of a power tool with interchangeable heads including soft wire brushes, carborundum heads and discs. Flexible carborundum discs are also available. Unfortunately this technique has to remove some of the surface to achieve cleaning; it is also notoriously difficult, even for an experienced operative, to avoid scouring flat surfaces with shallow depressions and leaving wavy arisses on external angles. Good-quality work is usually finished by hand rubbing to remove these imperfections, by which time the surface has, in effect, been re-dressed. In most situations it is a technique to avoid, unless there has been deep staining of a surface or a paint has to be removed which will not yield to solvents.

In recent years successful cleaning of flat surfaces has been achieved with a patent system feeding water and abrasive to small carborundum spinning heads, (sometimes described as 'polishing') although success is again directly related to skill levels and the resistance of the stone to scouring and scratching.

On a small scale, grinding down of thick encrusted dirt in detailed work can be safely and usefully achieved using pencil-sized tools with miniature interchangeable heads, driven by compressed air or electricity.

Compressed air and abrasive

The significant factors relating to compressed air abrasive cleaning are air pressure, nozzle size and type, type of abrasive, amount of flow of abrasive, skill of the operative, and supervision of the work. Each one must be selected and specified for the particular application.

An air abrasive system projects abrasive through a nozzle, in a stream of compressed air. The basic assembly of equipment is a compressor, a pot for the abrasive and air and abrasive delivery lines. Some types of system introduce water by running a twin hose to carry water to the end of the abrasive delivery line, discharging several small jets of water through a ring adaptor into the air and abrasive stream; other types mix the water, air and abrasive at source. Air pressures at the nozzle vary and there are different orifice sizes and nozzle patterns available. Small-scale and multi-faceted work has been successfully cleaned with air suction guns which use fine abrasives only. Typical pressures at the nozzle range are 3–14 kPa (20–100 psi) but can be as law as 7 psi.

During the cleaning of masonry, air pressures should be in the range of 1.5–6.0 kPa (10–40 psi).

Abrasives are selected according to the toughness of the dirt to be removed, but cost and safety factors will also have an influence. In general, 'round' abrasives such as shot, glass beads and some types of sand, hammer the surface; these are suitable where the soiling is hard and brittle over a fairly hard substrate such as a sulphated mortar film on granite. Angular abrasives, such as blasting grits, quartz sand or flint grit, have a cutting effect; these are suitable where a soft or resilient soiling covers the substrate. Sand is the cheapest abrasive, but the most hazardous. Harmful dust is always generated by dry blasting of sandstone and the operatives must have the protection of helmets supplied with filtered air and should wear full protective clothing. Other tough abrasives are non-siliceous grits, such as copper or iron slag, carborundum and aluminium oxide powders. These are all readily available throughout the UK. Increasing in use are 'safe' abrasives such as dolomite and olivine. For less demanding or more fragile surfaces, crushed egg or nut shells, minute glass beads or even talc may be used as the abrasive.

Nozzles need to be chosen carefully. Long venturi nozzles are most efficient and give an even particle spread over a greater impact area (at a constant distance from the surface) at any pressure than other types. They are suited to flat areas of consistent soiling conditions. Long and short straight nozzles, whilst less efficient, provide a more pencil-shaped blast which is suited to more detailed work. Angled nozzles are also available. It is also important to select nozzles which will stay the same shape throughout the job. The nozzle should also be able to provide an even, constant flow of abrasive to the work with the air/grit mix at any pressure being as lean as possible.

Operatives engaged in dry abrasive cleaning should wear protective clothing, the most important item being an 'air-line' helmet which, by maintaining a positive air pressure inside prevents the ingress of dust. The skill and experience of operatives is crucial to a good abrasive cleaning job.

Wet abrasive cleaning

The dry blast may be adapted to a *wet* process by use of a 'wet head' gun. There are several types of wet head which introduce water into the air and abrasive stream, either with single or multiple small jets. A mixture of water and abrasive tends to be less harsh than the dry abrasive, but this benefit is offset by the amount of slurry generated at the wall face. This slurry makes wet blasting unpopular with operatives because, even when they are properly attired, it tends to obscure the work by covering the plastic window in the helmet and by adhering to the wall face and collecting on ledges and in mouldings. The net result is that a light and dark mottled effect (known as 'gun-shading') and tenacious build-ups of dirt and slurry may be left behind.

However, there are also abrasive cleaning machines with nozzle grips which permit air, abrasive and water to be used in the following permutations: abrasive and air, abrasive and water, air only and water only. Selective use of these modes reduces sludge and visibility problems significantly.

An important advantage of the wet blast is that it reduces to a minimum the free dust which can be such a nuisance with a dry blast. Thus it is a pleasanter method for those using the premises, but the obvious temptation for some operatives to reduce or cut off the water during wet blasting must be guarded against. A wet-blasted facade should be well washed after completion, preferably with a high-pressure water lance, to remove dried films of slurry; build-ups of slurry on the scaffold and at ground level should be cleared away each day to avoid blocking gullies and surface water drains. Even though considerably less water is used during wet blasting than during washing, tarry 'drying out' stains must be expected where there have been heavy dirt deposits on porous blocks.

Dealing with noise and dust

The means of combating noise and dust have received a good deal of attention. The clouds of dust and abrasive can be contained by tightly screening the scaffold with reinforced translucent sheet and by sealing windows with tape, peelable plastic coating and sheet. But even when every precaution has been taken it would be unwise to exclude the possibility of infiltration of the finer dust particles. Although this is often an acceptable nuisance there will be special circumstances where this risk will rule out the method or at least dictate the use of a wet head. An abrasive pot is available with a drip dampener to cut down the dust nuisance. This can be moderately successful in dealing with the problem.

Noise may be a more serious problem, and again in certain circumstances may be a positive deterrent. There is no appreciable difference in noise levels between wet and dry blasting. The main problem is, of course, the noise of the gun and the impact of air and abrasive in the target area, rather than the background noise of the compressor which can be substantially muffled and sited to reduce the nuisance.

Evaluating abrasive cleaning

Factors which must be considered when the use of such a cleaning system is contemplated are:

- The relative hardness of the abrasive and the surface and the likely risk of damage
- The size of the particles of abrasive when appropriate. Coarse particles should be used for the preliminary cutting and fine for finishing
- The need for water to lubricate and cushion the impact effect of the abrasive and to reduce dust
- The risks of dust to the public and of penetration to sensitive areas of the building
- The available operative skills.

Problems associated with the air abrasive technique are as follows:

- The surfaces being cleaned and even the abrasive used can vary in hardness. Less resistant stones, or areas of the same stone, may be attacked by the air pressure and abrasive which cleaned a 'trial patch' without damage
- Grits of sand and flint contain free silica, as do sandstone and granite. Dusts generated during cleaning which involve these materials can cause long-term, irreversible lung damage to inadequately protected operatives
- The limited vision of an operative wearing a protective helmet is obscured in the immediate work area by dust, when large nozzles are used
- Dust can penetrate even small openings and cause nuisance or even damage furniture, fittings and machinery
- Compressor and air delivery noise can be a considerable nuisance to the occupants of a building which is being cleaned and may become intolerable in the immediate vicinity of cleaning

- Residual dust and spent abrasive will remain on the building giving an unnatural appearance, unless the cleaning is completed by washing down. This is most conveniently carried out by using a high-pressure, low-volume water lance, which does not involve any soaking of the building.

Advantages of the air abrasive technique may be listed as follows:

- Saturation of the building is avoided, even with 'wet-head' blast cleaning and lancing off, so that cleaning can usually proceed through the winter, even in cold climates
- There are few risks of water penetration and staining or efflorescence, although 'wet-head' blasting involves some risks
- On simple facades, the method is probably the fastest way to clean safely, assuming the necessary operative skills are available
- On small-scale, fragile detail, especially where there is a history of salt crystallization damage, small-scale air abrasive cleaning, if used skilfully, is safer than water cleaning or poulticing.

A range of equipment is now available which enables experienced operatives to clean a variety of surfaces safely, avoiding most of the mistakes of the past. The safety conscious water injection system can largely avoid hazardous dust during large-scale cleaning. Small air abrasive pistols and pencils, using dusts of 50–100 microns or finer can be used very effectively as supplementary tools to cleaning with water or chemicals, sometimes after the initial softening up.

The operative must be experienced and alert to changes which may become apparent in the surface on which he is working; adjustments in pressure or abrasive may be necessary, or a change of method, or work may have to be abandoned altogether if damage appears likely to a valuable surface. Work must never be hurried. Rushed air abrasive work, especially on flat surfaces, can result in a mottled 'gun-shading', which becomes apparent with subsequent weathering. Whether air abrasive cleaning is used wet or dry, the surface should always be finished with a water lance, in order to remove all dust and spent abrasive. On small-scale detail, hand sprays or air jets may be used.

Although air abrasive cleaning systems have become increasingly flexible they should not be expected to provide a total cleaning of all areas. Other cleaning methods may need to be used in conjunction. When a trial area of air abrasive cleaning is undertaken the operative should always start with the lowest pressure and the smaller, lighter abrasives. These factors can then be modified to determine the optimum work rate, abrasive type, size and pressure.

5.6 CHEMICAL CLEANING

Chemical cleaners are usually based on alkalis or acids. Most either contain soluble salts or react with stone to form soluble salts, and this means that they must be completely removed at the end of the cleaning operation. The only chemical cleaner known to leave no soluble salts in masonry is hydrofluoric acid, but because this is extremely dangerous in inexperienced hands its use should be left to firms employing trained operatives.

Hydrofluoric acid-based solutions (HF)

Hydrofluoric acid-based solutions are the chemical cleaning agents normally selected for cleaning sandstone and unpolished granite. Alternative systems should be considered for calcareous sandstones, which are acid-susceptible. They are also used to clean soft and heavily soiled brickwork which would not respond well to washing or would be irreparably damaged by abrasive cleaning. Only very weak HF at very short dwell times should be used on glazed or unglazed terracotta or faience. (See also Volume 2, Chapter 5, 'The repair and maintenance of brickwork' and Volume 2, Chapter 6, 'The repair and maintenance of terracotta and faience'.)

Hydrofluoric acid cleans sandstones by reacting with the silica which forms the main constituent of the stone. As the silica dissolves, the surface dirt bound to it is loosened and may be washed away. If there is a delay in washing off, some of the dissolved silica may be redeposited and it will show as a white bloom, or as white streaks from the joints. This redeposited silica is very difficult to remove. Although weathering generally will improve the appearance, it is of little help if the disfigurement is excessive. These deposits can only be removed by mechanical means, that is, by the use of an abrasive disc or an air abrasive unit, by water cutting, or by a further application of the acid. Since all these measures are damaging or risk further deposits of silica they are clearly undesirable in this context.

Sandstones which contain iron compounds present another problem. Hydrofluoric acid will attack the iron, forming soluble compounds which then migrate to the surface and form deep brown stains. The risk of staining can be considerably reduced, although not eliminated, by adding phosphoric acid to the hydrofluoric acid. Although attempts have been made to remove iron staining with subsequent applications of phosphoric acid and mixtures of phosphoric and hydrofluoric acids, the results have been disappointing.

Procedure for cleaning with HF

Procedures for cleaning a sandstone or unpolished granite building with hydrofluoric acid are as follows. All personnel must be experienced and be equipped with full face and head protection, heavy-duty gauntlets, waterproof boots and waterproof clothes and be familiar with the appropriate first-aid procedures. The local hospital should be informed that HF cleaning is taking place.

1 General procedures for protecting the building and sheeting the scaffolding must be taken, but particular care must be taken to protect contract personnel and the public from spillages or drift. First-aid boxes containing sodium gluconate gel must be kept on site. All scaffold tubes must be securely capped to avoid the trapping of acid or acid vapour. Glass should be coated with two applications of a latex masking paint (remember that some solvents are effective paint strippers!) and if the

window glass is particularly valuable, fitted with a polyethylene membrane and resin bonded ply templates. Templates alone, without the latex, should not be used

2 Use a proprietary, pre-diluted form of hydrofluoric acid: the concentration must be known and displayed on the container (between 2 per cent and 15 per cent). Do not store the industrial concentrate (which may be over 70 per cent) on site, or permit on-site dilution. Keep the acid in a secure store and adequately labelled

3 Pre-wet the area to be cleaned with clean water. A convenient way of achieving this is by using a low-volume, high-pressure water lance; remember that the objective is to provide a damp surface, on which the chemical will spread. If the surface is dry, or only superficially wet, the chemical will be absorbed, especially at mortar joints. Thorough wetting will limit the activity of the chemical to the soiled face

4 Apply the acid by brush to the damp surface, or by using a low-pressure spray. The application should be even and planned between architectural features (for example, cornice to plinth or between internal angles of buttresses). The coverage rate should be in the order of 1 litre per 3.7 square metres of surface area (1 gallon per 12–15 square yards). The contact period with the stone surface will vary, depending on the amount and type of soiling and on the ambient temperature. Proprietary hydrofluoric acid cleaners which carry an Agrément Board certificate (pH 1–1.5 and pH 3.5–3.8 are recommended) have a contact period of between 2 and 30 minutes, depending on the temperature. Repeated applications may be necessary. The cleaning material must never be allowed to dry on the surface

5 The importance of thorough washing off at the correct time has already been emphasized. This is achieved most efficiently with a high-pressure low-volume water lance (a pump producing, say 143 kPa (1000 psi) at 20 litres (4 gallons) per minute. A suitable technique is to hold the nozzle approximately 750 mm (30 in) away from the surface, while passing the lance to and fro, in sweeps of 750 mm (30 in). Rinsing for 4 minutes per square metre (square yard) is recommended as a minimum time, with extra attention being paid to water traps, such as cills and strings, or weathered joints. The rinsing water must not be allowed to accumulate on such traps as when the water evaporates, the acid concentration will increase and white deposits of colloidal silica may be left behind. Some proprietary materials contain a foaming agent, which identifies the presence of any residual acid.

6 The scaffolding boards must also be washed off thoroughly after each rinsing of the building and the scaffold tube capping checked.

7 Subsequent applications of the chemical must follow the same procedures, as described in (3) to (6). At least half an hour should elapse before a second application. If multiple applications are needed the method is probably the wrong one

First-aid treatment for HF burns

Burns must be washed immediately with copious amounts of clean water for at least one minute, followed by rubbing calcium gluconate gel into and around the burned area, with clean fingers. The gel should be rubbed in continuously until 15 minutes after the pain has subsided and hospital treatment must follow. If gel is not available, washing must continue until it is. Eyes which have been affected should be irrigated with isotonic saline or clean water for at least 10 minutes. Hospital treatment may involve injections of calcium gluconate into and under the burn and further treatment in the case of large or severe burns.

Ammonium bifluoride

Ammonium bifluoride is normally used for the cleaning of granite. Whilst this is in effect an HF-type clean, it is marginally dangerous and corrosive and adequate on most granite surfaces. The same safety precautions as for HF must apply. The cleaning action of ammonium bifluoride is similar to that of hydrofluoric acid but in addition the ammonia tends to emulsify grease and oils.

Other chemical cleaning agents

A limited range of other acid or alkali cleaning agents is available, but they all involve a risk of soluble salt residues.

Sodium hydroxide (caustic soda)

The most common alkaline cleaning agents are based on sodium hydroxide (caustic soda). Some contain surfactants and detergents to 'de-grease' a severely soiled surface before cleaning with hydrofluoric acid, in which case there is not likely to be a problem with residues.

Caustic alkali cleaners should not be used on porous brick as, even with adequate pre-wetting, more of the cleaner may be absorbed than can be satisfactorily washed off and soluble salts will be left behind. Caustic alkali cleaning of limestone should really be considered only as a last resort, when cleaning by other methods is not possible.

The same general procedures and safeguards as for cleaning with hydrofluoric acid should be followed, even though some hazards are less. In particular, the pre-wetting and thorough washing off are vital if staining and damage are to be avoided. Unfortunately, examples of such damage can easily be found, especially on the underside of window and door heads.

Alkaline cleaners compete with water washing, rather than blasting. Encrustations, heavy soiling, or indeed any soiling which requires more than two or three applications of the cleaner should be finished by another method. The soiled area should be first wetted, working from bottom to top to minimize the risk of streak staining and each application should be jetted off with clean water before the next is applied.

The advantages of this method over water are the considerable reduction in the quantities of water necessary to achieve a good result, reduction of the risk of staining and speed. Proper preparation of the site is essential. The low-volume, high-pressure water lances often used in jetting off chemical cleaners can force the chemical into open joints and cracks which have not been adequately sealed.

Hydrochloric acid
Hydrochloric acid is the other acid in common use for the removal of cementitious stains and deposits. Ten per cent hydrochloric acid applied to a pre-wetted surface will remove calcium carbonate; it is more likely to be used on brickwork and limestone than on sandstone.

First-aid treatment
First-aid treatment for sodium hydroxide ('caustic soda') and hydrochloric acid ('spirits of salts') is washing with copious amounts of clean water. Severe burns must be treated in hospital as soon as possible. When burns from hydrochloric acid have occurred, a magnesium oxide paste should be applied.

5.7 SPECIAL CLEANING SYSTEMS

Several special cleaning systems exist which are suited to specific problems and situations.

The Baker, or 'lime' method
The cleaning of limestone surface by lime poulticing as part of a total consolidation and protection programme was pioneered and developed by Professor Robert Baker. The system is fully described in Chapter 8.

Soaps
Grease, oil, tar and pitch will frequently respond well to scrubbing with warm water and a suitable soap, especially on marble or limestone. Slate, granite and even sandstone may sometimes be cleaned, or partially cleaned, by this method as well. Any of the deposit which can be lifted by a scraper or spatula should be removed first.

Powdered detergents must be avoided, because of the accumulative deposits of sodium salts which may build up, particularly in joints, after repeated applications (maintenance cleaning, for example).

Experience has shown that a proprietary methyl cyclohexyloleate, with a pH of 10.5–11.5, which is soluble in water and spirit, such as white spirit or trichlorethylene, is able to remove a wide range of soiling and has good penetrating effects into fine crazing and small cracks. This soap blend is non-foaming and remains active while it is on the surface (usually about 5 minutes). Thick deposits of greasy or oily dirt need to be worked on with bristle, synthetic fibre or soft brass wire brushes. Hand-warm water produces the optimum effect. After cleaning, all the soap should be rinsed away. Suitable, economic proportions are between 3–9 parts water/spirit to 1 part of soap. No particular safety precautions are needed, but the efficient de-greasing effect of the solution makes it advisable to wear protective gloves.

Alabaster, the surface of which is dissolved by washing with water, may be cleaned with the soap and white spirit and finished with white spirit alone on cotton swabs.

Poultice for limestone and marble

Heavily soiled limestone or marble can be cleaned with a poultice based on the chelating agent ethylene diamine-tetra-acetic acid (EDTA). The composition and method of application of this poultice are described in Chapter 7.

Clay packs

Certain clays make very efficient poultices for cleaning. A 50 micron attapulgite or sepiolite clay powder is added to enough clean water to produce a thick, sticky cream (the water should not be added to the clay, because a lumpy paste will be produced). This mixture may be applied to the surface of soiled limestone or marble on its own (no solvent is required) and covered with a thin polyethylene film. The poultice may be effective within a few days, but it may need to be left for several weeks. The contact period can only be determined by lifting the edge of the poultice and testing the tenacity of the surface dirt by gentle scrubbing. When a promising result is obtained, the complete poultice can be removed with a spatula and the surface scrubbed clean with bristle or soft, non-ferrous wire brushes. Used in this way, such poultices are a development of the more traditional 'wet-packs' composed of whiting, paper pulp or even bread and may be mixed with various solvents to lift stains.

Ultrasonic cleaning

The equipment used by dentists to descale teeth, developed some 25 years ago, is also used in conservation work, although normally in the workshop or laboratory. The tool has an ultrasonic vibrating head immersed in a spray of water, which flows around and through it. The vibration is transmitted into the water layer, creating movement, vibration and cavitation and thus cleans the surface.

Laser cleaning

Since 1972, laser radiation has been under investigation as a means of stone cleaning, specifically in the field of sculpture conservation. The attraction of the principle lies in the relative ease with which even encrusted dirt can be removed from the most fragile surfaces as a result of laser irradiation. A single pulse from the laser will clean a 25 mm (1 in) square area, a cleaning rate comparable with the air abrasive pencil, when the few seconds interval needed for the laser operative to redirect the beam is taken into account. The advantage of the method is the absence of any mechanical contact with the surface of the sculpture. Although this is an interesting and, in some ways, an exciting development in the cleaning of fragile surfaces, which shows much promise, it is unlikely that laser cleaning will be of practical value outside the conservation studio for many years.

5.8 APPROPRIATE CLEANING METHODS

The following methods are commonly found to be appropriate for the materials shown, provided the necessary skills are available.

Sandstone

- Air abrasive cleaning
- Hydrofluoric acid cleaning (4–15 per cent)

Limestone

- Washing
- Washing with neutral pH soap
- Air abrasive cleaning (often in combination with washing)
- Lime poulticing

Granite

- Polished – high-pressure water or warm water and neutral pH soap
- Unpolished – ammonium bifluoride (2–10 per cent)

Marble

- Washing
- Poulticing with attapulgite clay

Slate

- Washing with neutral pH soap

Terracotta and Faience

(See also Volume 2, Chapter 6)

- Washing with warm water and neutral pH soap
- Very weak HF (2–5 per cent) at very short dwell times (2–10 minutes)

Note: It is easy to destroy or damage glaze or fireskin with medium-strong HF, other chemicals and air abrasive tools. These methods should be avoided.

Brickwork

(See also Volume 2, Chapter 5 'Repair and maintenance of brickwork')

- Washing with pulse system
- Washing with warm water and neutral pH soap
- Washing with low-volume high-pressure water lances (sound, tough bricks only)
- Hydrofluoric acid (2–7 per cent maximum)
- Small air abrasive tools, rarely

Abrasive methods and brickwork

More damage has been done to brickwork by air abrasive cleaning than any other method, especially during the removal of paint. Air abrasive removal of paint, therefore, must only be carried out in exceptional circumstances by operatives of proven ability. Remember that a successful trial clean may not reveal brickwork which is typical of the whole wall and the operative must be sensitive to any change in the resistance of the substrate.

Wet methods and brickwork

Washing and chemical cleaning of all kinds involving the use of water is liable to produce efflorescent salts either from the bricks themselves or from the acid or alkali cleaning agents. It is essential therefore that scaffolding should not be struck before the cleaned wall has had a chance to dry out, so that the surface can be dry-brushed down. In circumstances where caustic alkali cleaners have been used and the cleaning is followed by efflorescence it is strongly recommended that after dry brushing the surface is re-wetted with a water lance and an absorbent clay plaster applied. Although this problem primarily relates to brickwork, similar precautions and similar remedial work, if necessary, should be taken with stone. Some of the most successful cleaning of brickwork has been carried out using weak hydrofluoric acid solutions with a contact period of between 5 and 10 minutes. Washing off must be very thorough but need not and should not involve saturation.

5.9 COMMON CLEANING PROBLEMS

Organic growth

There are many circumstances in which lichen and even some varieties of small plant may enhance the appearance of masonry without adverse effect. Other circumstances require a sterilizing treatment, for reasons of maintenance or appearance. Biological growths which should receive attention include unsightly algal slimes on vertical surfaces and especially on paving and certain acid-secreting lichens which cause the deterioration of some building materials, such as copper, zinc or lead sheet, or marble, limestone and glass.

This subject is covered in Chapter 2 'Control of organic growth'.

Paint removal

See also page 63 ('Old sulphated limewash') which follows.

The following treatments have been used successfully for the removal of paint from masonry surfaces:

- Conventional methylene chloride (paint stripper), in thick poultice under a plastic film
- Proprietary poultice-form stripper based on caustic soda (sodium hydroxide).

The poultice must be lifted off dry, not washed off, and the wall thoroughly washed after the poultice has been removed

- Hot air stripping (paint removal)
- Steam lance in association with strippers

Methylene chloride is the only suitable treatment for paint on terracotta and faience.

Removal of graffiti (aerosol paint)

Although most paints used for graffiti can usually be removed from the surface of victim masonry it is very difficult to remove pigment which has been carried into the pores by a solvent; sometimes the application of a solvent to remove the paint can drive the pigment more deeply into the stone, as for example, the application of cellulose thinners to freshly applied cellulose paint graffiti. Water-rinsable paint strippers, 1:5 solutions of water and trisodium phosphate and pastes of sodium hydroxide in clay have all been used with varying degrees of success. The technique is to leave the stripper in contact with the paint for long enough to cause softening and to enable scraping and brushing to take place successfully; the application of a thick layer is essential and a layer of thin plastic film may be necessary over the application. After the paint has been scraped off, the surface must be washed thoroughly, preferably in warm water and liquid neutral pH soap. Caustic strippers should be resorted to only when all else fails. Unfortunately, repeated attacks with paint and repeated removals build up an unsightly, patchy masonry surface.

On particularly valuable surfaces there is the possible refinement of picking out some of the stubborn pigment with an air abrasive pistol or pencil and a fine abrasive such as aluminium oxide crystals. In some situations the 'ghosting' left behind after cleaning may be further obscured by encouraging the development of lichen with 'organic washes' of animal dung in water or, better, of skimmed milk or plain yoghurt. The application of such material is, of course, not recommended on fragile or highly sensitive surfaces and will only occasionally be appropriate.

Anti-graffiti treatment

Areas liable to repeated graffiti attacks are sometimes treated with 'barrier application' to try to prevent the migration of paint into the surface pores of masonry and to facilitate removal. Such applications have attempted either to block the pores, or to cause temporary blocking by softening and swelling in the presence of moisture, or to line the pores with a water repellent coating.

Recent research by RTAS has shown that the pore-lining technique is the most successful. In a series of tests simulating repeated graffiti attacks on surfaces of varying porosity and permeability, either a single-pack, moisture-cured polyurethane or a two-pack polyurethane based on a colour stable isocyanate prepolymer proved to be the most promising. Cellulose paint was entirely removed by swabbing with MIBK (methyl iso butyl ketone).

No retreatment with either polyurethane 'barrier' system was found to be necessary after any of the paint stripping stages, unlike some currently available barrier treatments. In addition, the treatment is colour stable, need not change the appearance of the substrate, will inhibit the formation of organic growth, has good abrasion resistance (i.e. to repeated cleaning), allows simple and fast paint removal with non-caustic paint strippers, and allows the passage of moisture vapour. However, these barrier treatments must not be used on surfaces which are decaying. In such a situation it would be wiser to use a deeply penetrating alkoxysilane treatment which would serve to consolidate and give protection against paint attack. (See Chapter 9, 'Masonry consolidants'). Initial high cost of such treatments are likely to be offset by their persistent efficiency.

There are a number of proprietary graffiti-removers and graffiti-barriers on the market. They should be considered in the light of the above notes and examples of work cleaned should always be requested. Proprietary cleaners based on methylene chloride or rinsable solvent-based jellies and anti-graffiti coatings were tested during 1983–5 by the Conservation Laboratories of the Historic Monuments and Buildings branch of the DOE in Northern Ireland (Fry: reference 6). Tests werre carried out on proprietary barrier systems based on polyurethene resin or epoxy resin or acrylic emulsion. Only the acrylic emulsion treatment failed to resist normal weathering over the two-year exposure period, but the test report suggests that the useful life of any known anti-graffiti coating may be less than five years. There is no ideal solution for sites which are repeatedly attacked.

Removal of other deposits and stains

The removal of stains from masonry may sometimes be achieved with remarkable success, but more often only partial improvement is possible and in some circumstances nothing can be done. The latter categories have often led to painting over or scraping and grinding the surface away. The type of surface, the staining agent and the age of the stain affect response to cleaning attempts. Trial patches should always be made before attempting large-scale removal.

The following table shows methods and reagents which have been used for stain removal. In many cases an aid to cleaning may be provided by a *small* air abrasive tool used patiently at low pressure. Where poultices are used, the surfaces must be pre-wetted and all poultice material must be lifted off with plastic spatulas and placed directly into disposal bins or sacks, before thorough rinsing off with clean water. The longer a stain is left untreated, the more difficult complete removal becomes. Before cleaning, the means to avoid restaining should be decided upon; this may involve, for instance, the painting or removal of iron, or the regular treatment of bronze with lanolin and wax.

Copper stains

(Principally on limestone and marble)

Treatment
Repeated applications of the following poultice:

1 Add 70 g of ammonium chloride to 570 ml of concentrated ammonia. Add water to make the volume 1 litre. To 1 litre of this 'ammonia water' add 37 g of EDTA. Add attapulgite clay to form a soft paste
2 Pre-wet the surface with clean water, apply the paste and leave until dry
3 Remove the paste with a wooden or other non-metallic spatula
4 Rinse thoroughly with clean water
5 Repeat (2), (3) and (4) as often as required to lift or satisfactorily lighten the stain

Note: Traditional usage of ammonia direct or in a paste with whiting is only successful on light staining.

Iron stains

(Principally on limestone and marble)

Treatment

1 Make a mixture of glycerine (7 parts), sodium citrate (1 part) and warm water (6 parts)
2 Add attapulgite clay to the solution until a smooth paste is formed
3 Apply the paste to the stained surface and leave until dry
4 Remove the paste with a wooden or other non-metallic spatula
5 Repeat (3) and (1) as often as required to lift or satisfactorily lighten the stain

For very stubborn stains

1 Wet the surface with a solution of 1 part sodium citrate and 6 parts water
2 Apply an attapulgite wet pack, containing sodium hydrosulphite (sodium dithionite)
3 Lift off and follow by washing with copious amounts of clean water

Note: Some success has also been achieved using an amine complex of hydrocarboxilic acid in aqueous solution.

Iron stains may be removed from granite or sandstone by application of orthophosphoric acid, or from limestones, marbles and calcareous sandstones by solutions of sodium hydrosulphite. The stone must be pre-wetted and washed off thoroughly after application.

Smoke and soot

Treatment

- Scrub with a neutral pH detergent. The more stubborn patches can be pulled from the masonry pores using a poultice based on methyl chloroform. Another useful poultice contains trisodium phosphate (Calgon) and bleaching powder

Note: Methyl chloroform is often known as 1,1,1-trichloroethane. It has similar solvent properties to trichlorethylene and carbon tetrachloride whose usage is not recommended on safety grounds. Care must be taken with methyl chloroform although it is much less hazardous than the other chemical solvents mentioned. Ensure good ventilation and/or respiratory apparatus.

Asphalt, bitumen, tar and brown stains under soot

Treatment

- Where no surface damage will be caused, remove excess with a scraper or freeze with ice or dry ice and chip off with a small chisel. Then scrub with water and an emulsifying detergent and finally sponge or poultice with paraffin. Alternatively, wash with petrol. Naphtha is useful in breaking down bitumen

Chewing gum

Treatment

- Freeze with dry ice and pick off (small areas). For large areas or where gum is trodden in use pressure steam cleaning

Timber stains

(Brown or grey)

Treatment

- These stains are due to water spreading tannin or resin from the timber and can normally be removed by scrubbing with a 1:40 solution of oxalic acid in warm water

Note: Oxalic acid is extremely poisonous and must be stored and handled with proper care.

Water stains

Treatment

- Stains from running water can frequently be removed by scrubbing following wetting with a high-pressure mist. Alternatively, the treatment for lime mortar may need to be followed

Old sulphated limewash

Treatment

Old sulphated limewash in multiple applications is notoriously difficult to remove without causing damage to the surface below. During the nineteenth century considerable areas of limewashing were removed by scraping, leaving surfaces scarred and scratched. In some cases evidence of early colour schemes and designs were undoubtedly lost in the cleaning process. Only when it is genuinely desirable and archaeologically 'safe' and correct should limewashes be removed.

There are no acceptable fast methods of removing limewash. Poulticing with attapulgite or sepiolite clay kept damp under thin plastic film will soften the limewash. Scrubbing with phosphor bronze or bristle brushes and hot water will break down the surface. Hot water should be used if the limewash has an oil or tallow binder. Gentle cleaning with air abrasive tools can be used if the stone surface is not very weak. This system should be coupled with a small amount of water and the abrasive should be fine and sharp (angular). The equipment may be either suction or pressure at 1.5–3.0 kPa (12–20 psi).

The final clean will almost invariably have to be carried out by hand scrubbing. If possible, solvents should be avoided; if, however, residual traces remain the wall may be wetted with hand sprays, washed down with diluted hydrochloric acid (as for lime mortar, below) and rinsed with clean water on completion.

All these techniques require care and experience. In no circumstances are they to be used on subjects such as sculpture, church monuments or important detail, which are properly the province of the trained conservator.

Lime mortar

Treatment

- *Stone and clay bricks*: where possible, remove larger pieces of dirt with a scraper, then wash down with a diluted solution of hydrochloric acid (1:10 by volume). Prewetting and thorough rinsing off with clean water are essential
- *Calcium silicate bricks*: lightly abrade the surface using a brick of the same colour, then wash down with a solution of hydrochloric acid (1:20 by volume). The surface of some calcium silicate bricks may be damaged by acid, so this treatment should be used cautiously

Oil

Treatment

- Sponge or poultice with white spirit, or methyl chloroform (see note under 'Smoke and soot'). For calcium silicate bricks scrub with an oil-emulsifying detergent in water, allow to dry and poultice as above

Residual staining from pigeon gel

Treatment

- Sponge or poultice with methyl chloroform or methylene chloride (see note under 'Smoke and soot')

Grease, oil, food stains, hand marks

Treatment

- Poultice of trisodium phosphate (1 part), sodium perborate (1 part), talc (3 parts) in a hot neutral pH soap solution in water
- A useful proprietary de-greasing caustic alkali cleaner is available to break down greasy surface soiling, particularly on surfaces exposed to pollution from vehicles, and may be successfully used as a preliminary preparation for cleaning by other methods. Thorough removal is essential

5.10 THE LONG-TERM EFFECTS OF MASONRY CLEANING

During 1983–4, RTAS returned to about thirty sites to look at the condition of various building surfaces cleaned by different methods. All the buildings were cleaned by different methods. All the buildings were cleaned at least ten years earlier and all were situated in heavily trafficked urban areas in London, York, Edinburgh and Glasgow.

The inspections showed that:

- Although the specification was sometimes wrong the major problem related to inadequate levels of skill and poor or ignorant supervision on site
- Some cleaning methods inexpertly applied specifically encouraged re-soiling and had undoubtedly caused unnecessary damage initially
- Some buildings had been 'over-cleaned' probably because of client pressure to produce a 'pristine' appearance
- Water repellents were rarely successful as long-term dirt-inhibitors and could look grey or patchy.

The following table provides a summary of the observations.

Observations on masonry surfaces cleaned between ten and fifteen years ago

Surface	*System*	*Observations*	*Appearance and condition (typical after cleaning)*	
			Appearance	*Condition*
Limestone	Washing	Satisfactory to good, tarry stains fade, slow re-soiling	good	good

Observations on masonry surfaces cleaned between ten and fifteen years ago (*continued*)

Surface	*System*	*Observations*	*Appearance and condition (typical after cleaning)*	
Brickwork	Washing	Satisfactory to good, some persistent efflorescence, slow re-soiling	good	good
Limestone	Sand-blasting	Satisfactory – soiled areas 'blotchy', pitting noticeable on carved/moulded work, fast re-soiling	fair to poor	fair
Sandstone	Sand-blasting	Generally unsatisfactory, badly pitted surfaces, medium-fast re-soiling	fair to poor	poor
Terracotta	Sand-blasting	Unacceptable damage	poor	poor
Limestone	Chemical (alkali)	Some efflorescence and patchy staining	poor	poor
Sandstone	Chemical (acid)	Some iron staining, slights streaks of silica deposit	good	good
Brickwork	Chemical (acid)	Slight streaks from joints	good	good
Terracotta	Chemical (acid)	Streaks of silica, slow re-soiling generally on all surfaces	fair	fair to poor

REFERENCES

1 Ashurst, John, 'Cleaning Stone and Brick', Technical Pamphlet 4, SPAB, London, 1977.

2 Ashurst, John, 'The Cleaning and Treatment of Limestone by the "Lime Method"'

Monumentum, Autumn, 1984, pp 233–257.

3 Ashurst, John, Dimes, F., and Honeyborne, D, *The Conservation of Building and Decorative Stone*, Butterworths Scientific, 1988.

4 *Building Research Digests*, No 280 'Cleaning External Surfaces of Buildings'; No 177 'Decay and Conservation of Stone Masonry'; No 125 'Colourless Treatments for Masonry'.

5 British Standard Code of Practice, BS6270: Part 1: 1982 'Cleaning and Surface Repair of Buildings', BS5390: 1976 (1984) 'Code of Practice for Stone Masonry'.

6 Fry, Malcolm F, 'The Problems of Ornamental Stonework-Graffiti', *Stone Industries*, Jan/Feb 1985. (Malcolm F Fry, Conservation Laboratory, Historic Monuments and Branch, DOE (NI).)

See also the Technical Bibliography, Volume 5.

6 REMOVAL OF SOLUBLE SALTS FROM MASONRY ('DESALINATION')

6.1 DAMAGE TO MASONRY BY SALT CRYSTALLIZATION

The capillary movement of moisture through masonry is often associated with salt deposition which tends to be mainly concentrated at or close to the wall surfaces. The disruptive forces associated with the crystallization of these salts cause decay, usually seen as pitting, powdering and flaking of the masonry. Drying out of the walls associated with a damp-proofing treatment or the elimination of a ground water source may also lead to an increase in the amount of salt at or near the wall surfaces and deterioration may increase rather than diminish unless measures are taken to reduce the salt content of the masonry. Certain salts, particularly some chlorides, are hygroscopic ('water seeking') and can take up moisture directly from the atmosphere, and dampness and deterioration may persist even after rising damp originating from the base of the wall has been stopped. In these circumstances the salts must be removed or at least substantially reduced from the masonry if deterioration is to be controlled.

6.2 METHODS OF REMOVING SALTS

This chapter describes two methods of removing some of the soluble salts from decaying masonry; the use of clay poultices and the use of a sand:lime sacrificial render. These systems have been used where the sources of soluble salts have been rising damp, sea or estuarine sand in mortar, past flooding, storage of culinary salt, storage of gunpowder, storage of chemicals, human and animal urine and caustic alkali cleaning and weed-killing treatments. The techniques are mainly suitable for large, plain areas of masonry or simple architectural detail. They should not be used, except in a limited way by trained conservators, on delicate, damaged surfaces of carving or sculptures, nor should the process be used where the pre-wetting would create problems for plaster, painting or embedded wood or metal. Where poulticing or sacrificial rendering is considered to be appropriate it may need to be built into a long-term maintenance programme, perhaps every five or ten years, particularly if there is a persistent replenishment of soluble salts. In other situations, where the source of contamination has been removed, a single cycle of poultices may be sufficient to effect a long-term improvement.

Further information regarding rising damp in masonry can be found in Volume 2, Chapter 1 'Control of damp in masonry'.

6.3 TREATMENT OF SALT-CONTAMINATED MASONRY WITH A POULTICE

The form of 'deep washing' of masonry which is usually described as 'desalination' involves saturation and poulticing with an absorbent clay to try to reduce the level of potentially damaging soluble salts concentrated within the surface of decaying stone. To remove all soluble salts in the context of a building is, of course, impossible, but a significant reduction in the outer 100 mm may have the effect of stabilizing a previously friable surface or may prepare the way for a consolidant whose curing process would be seriously inhibited by a high concentration of, say, sodium chloride. A further use of absorbent clay packs is as temporary 'plaster' after the installation of a damp-proof course.

In principle the 'desalination' technique is very simple. A wall is saturated for several days by spraying with mists of clean water, until wetting has occurred for the full, or a considerable depth.* Fine sprays mounted on a boom delivering under 200 litres (18 gallons) per hour are sufficient to feed six spray heads covering an area seven metres square. The setting up of the sprays is designed to produce a consistent pattern of wetting.

The wetting period is determined by the construction of the wall and the porosity of the stone and mortar, but is likely to extend over three days and nights. Small areas may be persistently wetted from a back-pack sprayer, but this is labour intensive and tends to be less effective. In some situations it is sensible to carry out dry brushing before wetting, making sure the loose material is removed from the site. During the wetting process temporary gutters are required to collect the run-off from the wall surfaces and to conduct it to a gully or run-off point well away from the treatment wall base and any other walls. Heavy gauge polyethylene sheet, pvc guttering, timber battens and a syphon tube may be all that is necessary to provide an effective water catchment and drainage system. Sheeting should also be used to minimize splashing.

When the wetting process is complete the absorbent clay or diatomaceous earth (usually attapulgite or sepiolite clays, 50 mesh) is added to enough clean water to make a soft, sticky paste. Water must not be added to the clay or a lumpy, unworkable mix will be formed. The clay poultice can be mixed by hand or using a small mechanical mixer, depending on the quantity required. When

*During experimental work monitoring equipment is often set into the wall core, for example the work of the BRE at the Salt Tower in the Tower of London in 1974. See Reference 1.

Although 'de-salination' can never be total in the context of masonry, levels of salt contamination can be reduced to such a level that crystallization damage is significantly slowed down. This illustration shows hand application of an absorbent attapulgite clay pack to a decaying sandstone springer. Before the clay was applied action had been taken to reduce moisture movement in the wall and the stone had been irrigated with clean water. Three wetting and clay-packing cycles were needed, allowing the salts in solution to dry out into the clay pack.

free of lumps the poultice is plastered onto the wet treatment wall in a single layer 20–25 mm (up to 1 in) thick using a plasterer's float or broad trowel. A 50 kg bag of clay will cover approximately three square metres. In its freshly mixed state the clay has very good adhesion and can be levelled reasonably accurately, even by a relatively inexperienced person. An important part of the technique is the ironing on to ensure good contact at all points. To help the clay keep its bond for as long as possible a light-gauge galvanized wire mesh is pressed into it and tacked carefully into joints with galvanized staples. Any springiness in the mesh can be reduced by localized cutting with wire snips, pressing the cut ends into the clay. In some situations, especially where the wall surface is heavily contoured as in corework, the overall adhesion of the clay will be assisted by cutting strips of open weave hessian soaked in a runny slurry of attapulgite clay and pressing these into the poultice. These strips, approximately 75 mm wide, may be used alone or with wire. Wire is essential on a large area of flat surface where the weight of the clay tends to induce pulling away from the wall.

When the treatment wall is fully plastered it must be protected from direct sun or rain or, if internal, from any heat source which will produce rapid drying. Externally, a ventilated space can most easily be set up with a tarpaulin or reinforced plastic sheet as a tent.

As the poultice dries out it draws salt-laden water from the masonry. Water evaporating from the clay face leaves behind salt crystals which can usually be seen in the form of efflorescences on the clay or wire. Drying conditions and the thickness of the wall dictate the contact time which varies very considerably from a few days to weeks. One month is not unusual for drying out, during which the clay lightens in colour, cracks, shrinks and detaches from the wall. At this stage the staples are withdrawn with pliers, and the bulk of the clay may be rolled up on its wire reinforcement. The spent clay should be put at once into plastic sacks or otherwise removed safely from the site. Small amounts of clay still adhering may be brushed off the wall with a stiff bristle brush. These sweepings must also be removed from the site.

The cycle of wetting and poulticing may need to be repeated several times to reduce the salts to an acceptable level. Salt sampling and analysis may need to be carried out to determine the levels present. Clay poultice 'desalination' is a lengthy process but it does not require a lot of supervision, expensive equipment or highly skilled personnel. It is best scheduled in with other works on or near the site.

Clay poultice 'desalination' has mostly been used on stone walling. On brickwork or rubble, where there are plenty of 'keys' for the poultice, special care needs to be taken when brushing down to remove all traces of dry clay from the joints. Any masonry which has been subject to extended periods of salt crystallization may well require pointing at the completion of the desalination treatment.

6.4 TREATMENT OF SALT-CONTAMINATED MASONRY WITH A SACRIFICIAL RENDER

Where it is not possible to remove excessive amounts of salts with the poulticing technique, the application of a porous sacrificial render may provide a more practical method of overcoming the problem. A porous render is applied to the wall and evaporation of moisture from the wall results in soluble salts being transferred from the masonry to the render. The render will deteriorate with time and may require renewal, but the masonry will be protected against continued decay. A sacrificial render can be used either to reduce the salt content of a wall where rising damp treatment will also be carried out, or it can be used to protect a wall against salt attack where rising damp cannot be prevented.

The wall is first wetted and a render of 1 part slaked and screened lime putty to 4 parts fine sand is applied at least 12 mm thick to both sides (if possible) to a height 50 mm above the salt crystallization/evaporation zone. The render should not be overworked with a trowel as optimum moisture evaporation and salt transfer will be obtained when the render has an open texture and a rough finish to increase the surface area. A practical and visually pleasing way of achieving this is to scrape the surface down after rendering with the fine-toothed edge of a hacksaw blade. This is carried out after the surface has begun to stiffen.

As salts transfer to the render and crystallize there, the render will begin to break down. Salt-contaminated render deposits at the base of a wall should be collected frequently. Where a contamination is severe, the application of only one render coat may be insufficient to reduce the salt content to a safe level and further treatment will be required. The remains of the first coat should be carefully removed, the wall re-wetted and the second coat applied.

Sacrificial sand:lime renders are a relatively slow method of masonry desalination. A period of several months may be required depending on the level of salt and the amount of evaporation. Most success has been achieved on walls where rising damp was still present, before any damp proofing installation was carried out. The process is, however, inexpensive and easy to undertake. The method was developed in Australia (see Reference 2) where it has been used successfully on sandstone and brickwork.

6.5 DETERMINING SALT LEVELS

If a more than superficial assessment is to be made of poulticing or sacrificial rendering, the level of salt within the masonry should be determined before, during and after the programme of work. Initially it may also be important to determine the types of salts and their hygroscopicity. The services of a laboratory will be required here.

Samples should be taken at depths of 0–25 mm, 50–75 mm and 75–100 mm, within the zone of evaporation and deterioration, usually about 900 mm (36 in) from ground level. After a clay poultice has been removed some drying out and migration of salts will continue, therefore, the wall should be allowed to dry before further samples are taken for salt analysis.

REFERENCES

1 Bowley, M J, *Desalination of Stone: A Case Study*, BRE Current Paper Series CP 46/75, HMSO, Garston, England, 1975.
2 Heiman, J L, *The Treatment of Salt-contaminated Masonry with a Sacrificial Render*, Technical Record 471, Commonwealth Experimental Building Station, Ryde, NSW, May 1981.

7 CLEANING MARBLE

7.1 CLEANING OF MARBLE

This chapter describes the cleaning and surface treatment of architectual marble. Several of the techniques are those used in sculpture conservation, and it is important that valuable marble pieces, especially those in poor condition, should be treated by a trained sculpture conservator. The chapter does not, however, attempt to cover all aspects of marble sculpture conservation.

The surface of marble may be damaged or disfigured in many ways. An originally polished surface may become roughened by exposure to a polluted atmosphere (acid attack and repeated crystallization of soluble salts), or by contact with acid secretions from algae and lichens or wood. Staining commonly results from contact with human hands, iron, copper, bronze, mortar (especially cement mortar) oils, tobacco smoke and shellac.

The methods described below may be used to improve the appearance of the marble. A small area of about 50 mm square should be cleaned first to observe the effects of the chosen system/s. It must be remembered that all acids are potentially dangerous to marble surfaces, and that their use must be very strictly controlled.

7.2 CLEANING SYSTEMS

Before any cleaning is commenced a careful investigation should be made to determine whether there are any remnants of pigment. Spot fixing of these with acrylic silane may be necessary prior to cleaning.

General cleaning

Marble may be washed with clean water in the same way as limestone if the soiling is water-soluble, sooty material. Cleaning can proceed more easily if progressively hotter water is used. The quantity of water should be kept to a minimum. On small areas it may be convenient to soften the dirt by hand

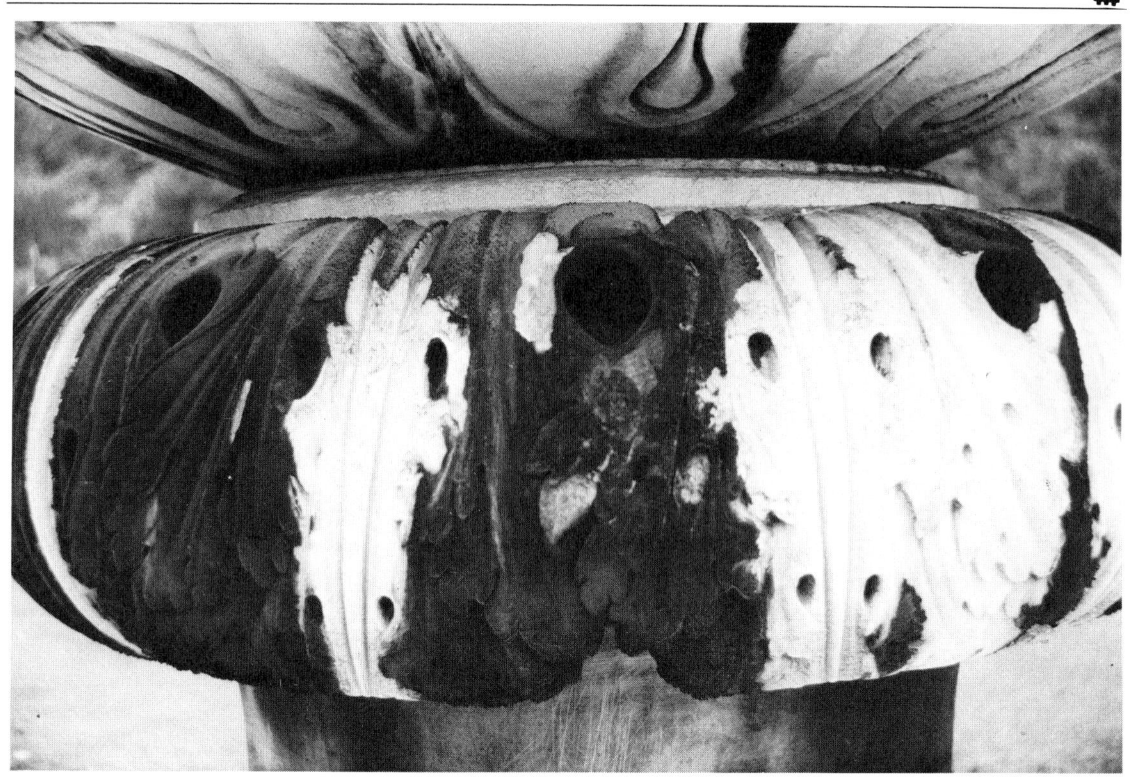

In an aggressive atmospheric environment such as London's, marble is always at risk. Acid attack will begin to etch marble surfaces within a few months and progressive deterioration is evidenced by sugary rainwashed zones and blistering sulphate skins, sometimes with heavy encrustations, as shown in this illustration. Solvent cleaning packs and small air abrasive tools are often the most effective and safest method of cleaning, sometimes with secondary packs to remove deep-seated stains.

spraying and by applications of wet absorbent clay such as attapulgite or sepiolite, trowelled onto the surface and secured with bandages of scrim. When the clay is lifted, light scrubbing following the direction of the sculptor's/carver's tools with a stiff nylon or bristle brush and further hand spraying may be sufficient to release the softened dirt. Valuable pieces should be cleaned with applications of absorptive clay and de-ionized water, and the loosened dirt removed with cotton wool swabs. A neutral pH soap and water (1:1) or white spirit and water (1:1) should be used to remove more intransigent general grime.

Poultices for heavily soiled marble: the Mora poultice

Heavily soiled marble can be cleaned with a poultice based on the chelating agent ethylene diamine-tetra-acetic acid (EDTA). The poultice facilitates the dissolution of calcium and iron salts by complex formation. In common with other acids, EDTA must not be used alone on marble. The following proportions have been used with considerable success to clean moderately soiled marble.

In 1000 ml water
- 60 g ammonium bicarbonate
- 60 g sodium bicarbonate
- 25 g EDTA
- 10 g surfactant disinfectant (such as 'Cetavlon')
- 60 g carboxymethyl cellulose (a cellulose-based wallpaper paste is a readily available substitute)

The ammonium and sodium bicarbonate give a slightly basic mixture of pH 7.5 and facilitate the dissolution of some salts.

The above poultice is known as the Mora poultice and was developed by Paulo and Laura Mora at the Istituto del Restauro in Rome.

The poultice, in the form of a clear jelly, is applied to a pre-wetted surface by spatula or by brush as a 3–4 mm thickness and is covered at once with a thin polyethylene film to prevent drying out. The film is of utmost importance as the cellulose body of the poultice is most difficult to remove if it dries and hardens. Contact period may be twenty-four hours and intermediate lifting and reapplication may be necessary. After cleaning and removal of all poultice material by the use of small trowels or spatulas the surface should be washed thoroughly with clean water.

The advantages and attractions of this system are principally that it is safe chemically and avoids any excessive use of abrasion or water; it cannot, however, be applied to friable or flaking surfaces, any more than any other poultice, without removing surface material. Surprisingly large areas can be cleaned relatively economically when the surface is not too detailed.

Stain removal

Marble is frequently stained by iron, bronze, copper, oil and grease. Washing or poulticing with water will not improve the appearance and it may not be possible to remove the stains completely, particularly if they have been there a long time. Stains from oils which have penetrated deeply and oxidized in the marble are particularly stubborn. Where possible, to determine the type of stain, careful testing on a discreet section of the marble should be carried out.

The following poultice packs are traditional and can usually be relied on to lighten if not to remove the stain. The main function of a poultice is to draw dissolved staining matter out of the marble.

Iron stains

- A solution of 1 part sodium citrate and 6 parts of water is added to an equal volume of glycerin. Attapulgite clay (traditionally, whiting) is added to the solution until a smooth paste is formed, which is then applied to the surface with a small trowel or putty knife. When dry the paste should be

removed with a wooden or other non-metallic spatula. The paste may need to be renewed several times (1)

- Very stubborn stains may require an alternative treatment using a thin layer of sodium hydrosulphite (sodium dithionite) crystals held in contact by an absorptive clay body. The marble should first be wetted with a solution of 1 part sodium citrate: 6 parts water. Lift off the poultice and rinse off thoroughly with clean water (1)
- Iron staining will also respond well to an amine complex of hydro-oxycarbolic acid in an acqueous, thixotropic form, or to a poultice of EDTA (page 74)
- Very light staining may be removed by a poultice including oxalic acid, based on 1 part acid powder to 10 parts water by weight, in a clay or acid-free paper pulp body

Bronze or copper stains

- Mix dry 1 part of ammonium chloride with 4 parts powdered talc, and add a 10 per cent solution of ammonia water. Pre-wet the marble with clean water, and apply the paste with a putty knife and leave until dry. Remove with a wooden spatula. Rinse off very thoroughly with clean water. The process can be repeated as often as required to lift or satisfactorily lighten the stain (1)

Oil stains

- Mix 1 part of acetone and 1 part of amyl acetate; soak a non-dyed cotton flannel in the solution and place the flannel over the stain. Alternatively, clay may be used to form a pack. Keep the flannel or the clay on the stain for three days under a thin film of plastic. Alternatively, a thixotropic form of methylene chloride may be applied as a pack. Light oil staining can sometimes be removed with xylenes, also in poultice form. (The general use of xylene is not recommended on health and safety grounds) (1 and 2)

Smoke

- First wash the surface with a neutral pH soap and water. Smoke stains have been successfully removed with the EDTA poultice (page 74). More stubborn patches may require treatment with a clay poultice based on trichlorethylene. (General use not recommended on health and safety grounds)

Bleaching of stains

- Stains which do not respond easily to any of the above treatments may need to be treated with a mild bleaching agent which can be applied by brush or held in place with a poultice. Hydrogen peroxide, 3 per cent in water, activated with a drop of ammonia can be an effective agent for bleaching out certain classes of stains

Stain removal recipes of the 1920s

The following recipes date from the 1920s (reference 1) and are still sometimes used in the masonry trade.

Ink stains

- Ink stains were traditionally removed by a poultice of whiting mixed to a thick paste with a strong solution of sodium perborate in hot water. The poultice was applied in a layer 6 mm ($\frac{1}{4}$ in) thick, left until dry and repeated if some of the blue colour remained. Synthetic dye inks were removed by the above method as well as a whiting poultice with ammonia

Tobacco stains

- Tobacco stains were removed by preparing a poultice of fine powder such as talc or whiting. The powder was stirred into hot water until a thick mortar-like consistency was obtained and then mixed thoroughly for several minutes. It was applied to the stained marble in a layer 12 mm ($\frac{1}{2}$ in) thick, left until dry and removed with a wooden scraper. Two or more applications were usually necessary
- An alternative method was to apply a poultice of soap and soda. A solution of 25 mm^3 (1 cubic inch) of soap dissolved in 1.2 litre (one quart) of hot water was combined equally with a second solution of sodium carbonate, prepared by dissolving four tablespoonfuls of washing soda in 1.2 litres (one quart) of water. The poultice was made by mixing a portion of the soap and soda solution with a powdered talc or whiting, applied to the stain and left to dry

Fire stains

- The smoke or pitch from burning wood was traditionally removed by soaking an undyed flannel cloth in a solution of trisodium phosphate and chlorinated lime, pressing this firmly against the marble and covering it with a piece of glass or marble. (Chlorinated lime is better known as bleaching powder, and is still available under that name)

Perspiration stains

- Perspiration stains were traditionally removed by the fire stains method

Coffee stains

- These stains were removed by placing a cloth saturated in a solution of 1 part glycerin and 4 parts water over the area

Removal of paint

A thixotropic paste of methylene chloride can be applied. Only in the case of stubborn paint surfaces might it be desirable to leave the paste under a plastic film. The paste and paint should be removed with a wooden spatula and the surfaces then thoroughly washed.

Removal of organic growth
Apply a quaternary ammonium fungicide as described for limestone and sandstone (see Chapter 2).

7.3 ACCRETIONS ON MARBLE

Marble which has been buried may be extensively stained and have on its surface deposits of sand and dirt, some of which may be cemented by calcareous material either precipitated from ground waters or washed out of the body of the marble. Similarly, calcareous deposits may form on marble elements of a fountain. These deposits can be almost as hard as the body of the marble itself and their removal is always a difficult matter.

Loosely adhered accretions may be freed by rinsing with de-ionized water to remove soluble components and a minimum of mechanical effort. Accretions which are more firmly adhered can be removed with a sharp scalpel and/or an air abrasive pencil.

7.4 REPOLISHING MARBLE

The repolishing of roughened surfaces can be carried out with a very mildly abrasive putty powder and chamois leather, followed by rubbing in a micro-crystalline wax to achieve a soft shine. Apart from the visual improvement, this treatment significantly inhibits re-soiling and further deterioration.

7.5 CONSOLIDANTS

Preservative treatments are described separately in Chapter 9, this volume, 'Masonry consolidants'. Adequate depth of penetration is always likely to be a problem with marble.

REFERENCES

1 Kessler, D W, *A Study of Problems Relating to the Maintenance of Interior Marble*, Technological Papers of the US Bureau of Standards No 350, US Government Printing Office, Washington, 1927.
2 Rinnie, D, *The Conservation of Ancient Marble*, The J Paul Getty Museum, 1976.

8 THE CLEANING AND TREATMENT OF LIMESTONE BY THE 'LIME METHOD'*

8.1 DEFINITION

The 'lime method' refers to a process of cleaning and treating limestone which was substantially developed by Professor Robert Baker during work under his direction in Oxford, Bath and Wells from the 1950s onwards.

The lime method is best described in the following sequence. This description refers to the use of the method in the context of sculpture. The method is applicable and has been successfully used on architectural detail and ashlar surfaces. See Volume 3, Chapter 8, 'Cleaning and consolidation of the chapel plaster at Cowdray House ruins', and Chapter 11 this volume, 'Case study: the consolidation of Clunch'.

8.2 SURVEY

In common with other proposed schemes of treatment, any activity should be preceded by a careful survey of the general and detailed environmental influences and the condition of the subject. If the subject is very important and the conditions complex, the conservator may need to call in the services of other specialists. For instance, it may be advisable for an art historian to make an assessment of sculpture before it is touched, or a consultant to look at the conservation of fragmentary polychrome. What is of importance to the conservator is thorough familiarity with the subject before work commences; for this reason, the survey should always be part of the conservation exercise and not carried out by someone other than the conservator.

*Parts of this chapter first appeared in *Monumentum*, 1984, summer edition. For the full text, see *The Conservation of Building and Decorative Stone*, Ashurst J, Dimes, F G, Honeyborne, D B (Butterworth Scientific 1988).

Although developed for the treatment of surfaces of very high intrinsic value, the lime method is, in principle, a very simple and successful way of cleaning, consolidating and preserving any limestone detail. The sculptured frieze in the illustration was washed, consolidated where necessary with limewater, and treated with suitably coloured lime-casein shelter coats. The shelter coating will require renewal, perhaps in eight to ten years, but even if this does not happen the treatment will have achieved some slowing down of the weathering processes.

The completed survey must include adequate photographs and drawings, to record and explain all the information listed above, in addition to notes and appropriate measurements.

8.3 STRUCTURAL REPAIR

Although structural repairs are sometimes the first operation, most do not take place until the cleaning has been completed. They are not described here in detail, as they do not relate specifically to the 'lime method', but they are likely to include the removal and replacement of iron dowels and reinforcement with threaded stainless steel. All new dowels and pins are set well below the surface of the stone and plugged with a carefully matched repair mortar.

The traditional method of cleaning associated with the lime method is by hot lime poultice. In the early days of experiment at Wells, the lime was slaked against the surface of the stone. Hot lime is still used, but is now applied by gloved hand and trowel, pressing the putty well into the surface of the pre-wetted stone. When a thick plaster has been applied, it is bound with scrim and wet sacking or underfelt secured with string. Finally, a heavy-duty polyethylene sheet is tied loosely in position. From time to time, over a period of two to three weeks, the polyethylene is lifted and the felt/sacking surface sprayed with water to ensure that the poultice remains damp and soft. This is essential to ensure that no drying out occurs which would render the lime useless as a poultice, or bind it to the surface of the stone.

When the packaging is finally removed, the lime is carefully lifted off in small areas at a time with spatulas or small trowels, taking with it some of the dirt from the contact surface. Water sprays are used to assist in the removal of the lime and to further soften the dirt. In common with most other poultices, not very much dirt actually detaches with the poultice material; the softened deposit must be worked at with hand sprays, dental picks and small toothbrushes or stencil brushes, to achieve a clean, or relatively clean, surface. The scrubbing stage may be long and laborious; added to two or three weeks' poulticing a life-size figure may well extend to a month or six weeks before any repair work is undertaken. It should be noted that an air abrasive unit may preserve more of the polychromy than a lime poultice would have done, particularly in view of the risks involved with water during removal of the poultice.

Wet poulticing combined with careful mechanical cleaning (dental tools and brushes or air abrasive) seems to be the best option for safety (of the surfaces and operatives) and economy. Whether or not the poultice should be lime, hot or cold, or attapulgite clay, or some other medium is still open to debate. Any 'wet pack' which can remain in intimate contact with the stone without drying out, or adhering to the surface, will have a softening effect on dirt and make it more responsive to gentle washing and brushing.

8.4 REMOVAL OF OLD FILLINGS

The cleaning processes reveal the full extent of the damage due to weathering and decay and the amount of cementitious filling to spalls and cracks which have been carried out in the past. These fillings are always unsuitable as a visual match for the stone, but, more seriously, their dense, impervious nature encourages moisture and salt concentrations around them, extending the area of decay still further. Removal of these old fillings is essential and requires slow and careful work. Once removed and their cavities recut and new cavities formed where newer spalls and splits had developed, all loose dust and debris must be removed by flushing with clean water. If organic growth has been present, flooding with a biocide is necessary to provide a clean, sterile surface for the new mortar filling.

8.5 CONSOLIDATION BY LIMEWATER

The cleaned surfaces with open cavities are next treated with limewater to attempt to consolidate the more friable areas.

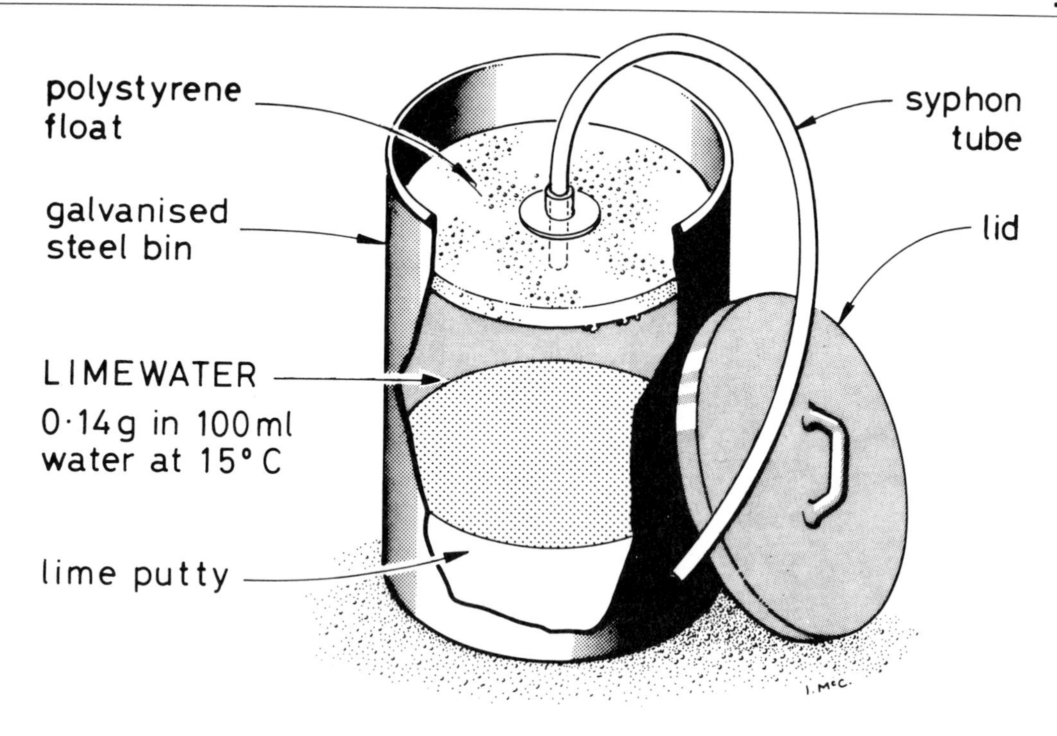

Figure 8.1 Limewater storage

Limewater contains small quantities of calcium hydroxide (0.14 g in 100 ml of water at 15°C). Traditionally, it is siphoned from the slaking tank after lime has been slaked in excess of water and after all slaking has ceased and the water is clear. Usual practice now is to stir lime putty into a container of water and leave it to stand until the water is clear. It is important that the limewater is protected from the air, otherwise it will carbonate and become ineffective. A number of different methods have been used to achieve this in a practical way; the most recent development is the covering of the surface of the limewater in its container with a float of polystyrene sheet stock, pierced only by a siphon tube fitted with a filter.

The limewater is drawn off when required by a hand pump into spray bottles or directly to a lance with a control valve and adjustable nozzle, checking from time to time that the water has not accidentally become clouded through disturbance of the lime in the bottom of the bin. Any cloudy water is rejected and the water allowed to stand until it is clear again. Approximately forty applications of limewater are flooded onto the surface of the limestone over a period of several days; application can continue as long as the surface will absorb, but excess limewater should not be allowed to lie on the surface of the stone and is removed by sponges which are then squeezed out in clean water.

Consolidation effects have been reported over many years as a result of multiple applications of limewater to lime plaster, Doulting, Bath, Clunch,

Barnack, Beer, Salcombe and Chilmark limestones, although it has to be said that attempts to record or quantify the phenomenon have met with a disappointing lack of success.

8.6 SURFACE REPAIR

The consolidation treatment is followed by the placing of mortar repairs, the stage of work in the 'lime method' where perhaps the greatest skill and the most experience are needed. Certainly the mortar repair is the core of the method and, when well executed, is the work which evokes the greatest admiration.

All mortar repairs are based on lime; no Portland cement of any kind is used. If a weak hydraulic mortar is needed, then a small addition of 'HTI' powder (a white refractory brick powder – 'high temperature insulation') is used as a pozzolana. All limes are of a 'high calcium', non-hydraulic type.

Lime is brought to the site after burning and is slaked as soon as possible in a suitable tank by adding it to water, raking and hoeing it through, until all visible reaction has ceased. An excess of water is used, so that the soft mass of lime putty formed during slaking is kept well covered. The lime putty must be left in its tank under water for as long as possible. The absolute minimum should be one week to ensure that all slaking is finished, but any days, weeks, months or even years that can be added to this period can be looked on as a bonus, especially if the lime putty can be mixed and stored in wet, airtight conditions with the aggregates. Of course, it will never 'set' or harden too much if it is kept from the air and, even if it has stiffened, it can easily be 'knocked-up' again when needed without the addition of any water; with sufficient working it will soon become a soft gelatinous mass again. Pozzolanic additives must only be added just before use, and then mixed in very thoroughly indeed.

Aggregates are selected and graded for colour and function. Considerable time is spent in their selection and many sands and crushed stones will be tried in the process of finding the right combination. Stone pieces are crushed by a hammer or roller on a concrete slab, or even in a corn grinder and then carefully sieved and graded for storing in a 'bank'.

The mortars have a number of different functions to fulfil. They are all likely to be a combination of lime and the same aggregates, but the lime:aggregate proportions may vary and so will the size of aggregate; only some functions require a pozzolanic additive. Basic proportions may be summarized as follows:

	Lime : *Aggregate*
Repair mortar	1 : 2
Adhesive mortar (for fixing spalls)	1 : 1
Grouting mortar (for crack filling)	1 : 1½
Shelter coating	1 : 2-3

Ten per cent pozzolanic additive ('HTI' powder) is included in the basic aggregate proportion in the adhesive and grouting mortar, but a lower percentage in the repair and shelter coat mortars. The aggregates tend to become finer towards the bottom of this list and are always very fine indeed for shelter coating. Some examples of aggregate sizes related to mortar function are given below.

Mortar function	*Lime*	*Aggregates BS sieve sizes* 1.18 mm	600 µm	400 µm	300 µm	*Pozzolanic additive* 600 µm	300 µm
Mortar repair	3	1½	1½	¾	¾	½	–
Mortar repair	3	3	2	1	–	½	–
Adhesive mortar	6	–	–	–	6	1½	–
Adhesive mortar	6	–	1	1	4	½	1
Grouting mortar	3¼	3	–	1	1	½	–
	2	–	½	1¼	–	–	¾
Shelter coat	3	–	–	–	8	–	–
Shelter coat	3	–	–	2½	4½	–	–

The lime putty is always screened through a 1.18 mm mesh after slaking. Further screening takes place according to the function of the mortar.

The final colour of the repair is dependent on the selection and blending of the aggregates and the proportion of lime used, the method of placing the repair and the rate of drying out. Minor variations in colour continue to take place indefinitely just as the colour of a stone surface will continue to respond to variations in humidity. Successful 'instant effects' are not necessarily the most satisfactory after a period of weathering and only considerable experience can design for this and select the mortar constituents accordingly.

When the conservator goes onto the scaffold to place the repair mortar a number of pre-mixed mortars, sometimes as many as thirty for two stone types, will have been prepared in separate plastic tubs, covered with a piece of wet cloth and the tools and materials necessary for the operation will be conveniently laid out on a board ready for use. These will include hand-spray bottles full of water, cotton wool packs, a small trowel, dental picks and plugging tools, spatulas, two or three small bristle brushes and rubber gloves. The following sequence of working is typical.

As a general rule no modelling is carried out in the repair mortar. Its role is to fill cavities and cracks and to provide a weak, porous capping to vulnerable, friable areas; it is designed to draw moisture and therefore soluble salts to itself and finally to fail before any further stone is lost. Ideally it will then be replaced.

1 Cavities and cracks are flushed out again with water from the hand sprays to avoid an otherwise dry stone surface de-watering the repair as it is pressed into position. The surface should be damp without water actually shining on the surface

2 Deep cavities are treated at the back with a slurry of repair mortar followed by a filling into which small pieces of limestone are inserted to reduce the thickness which needs to be built up in fine repair mortar

3 A thin slurry of repair mortar containing 'HTI' powder is brushed into the cavity or fracture to provide an additional key for the repair

4 After one or two hours, when the slurry has dried, the cavity is wetted up again and the first repair mortar is kneaded and pushed into place with the fingers, exerting as much pressure as possible. With few exceptions, not more than 5–6 mm should be pressed in at one time. Dental plugging tools and spatulas are used to assist in the filling. Throughout the entire sequence compaction of the amalgam by pressure is absolutely essential to achieve good adhesion and minimum shrinkage

5 As each filling is completed precautions must be taken to avoid rapid drying out by protecting the area from direct sunlight or strong draughts. When dry, the cavity must again be wetted and stage (4) repeated until the cavity has been filled completely. Overfilling is a useful aid to compaction and surplus mortar can be trimmed off with a spatula to the desired profile on completion. A texture matching the stone can be achieved with a dry sponge (taking care not to press hard and absorb moisture from the repair), hessian pads, stencil brushes and purpose-made plastic scrapers

To ensure slow drying, wet cotton wool packs are laid over the finished repair and left in position for as long as is thought necessary.

8.7 SHELTER COATING

The final stage of the work is to apply a thin surface coating to all the cleaned and repaired stone. This is intended to slow down the effects of weathering on the surviving surfaces by providing a sacrificial layer which may be removed by direct rainfall or disruption by salt crystallization activity associated with wetting and drying cycles. In the case of stones which were once covered with gesso and coloured with tempera the shelter coat may be seen as a substitute protection. The shelter coat is of similar or the same composition as the repair mortars, but the aggregate to lime proportion is slightly higher and sand and stone dust are crushed more finely. Water is added to the fine lime and aggregate mix until a consistency of thin cream is reached and thorough mixing continued for 20 to 30 minutes. Casein and formalin are added at the end of the mixing.

Careful colour-matching of cleaned, weathered stone precedes the full application. This matching can be carried out on a separate piece of the same stone in similar condition, but it is better to lay the samples on the stone itself or on, for instance, an adjacent moulding. Considerable skill is required in colour matching, as in matching repair mortars. All trial colours must be completely dry before a decision can be made about its accuracy. Sometimes a hot air-blower may be used to hasten the drying of the trial colours.

The surface is prepared for application by careful but thorough spraying with water. Spraying is carried out with hand bottles until water begins to sit on the surface and is no longer absorbed into the stone. At this stage, as soon as the water has ceased to glisten on the surface, the shelter coat can be laid on with a soft bristle brush. A second, short-haired (or worn) bristle brush is used to work the shelter coat into the texture of the stone. Traditionally, pads of hessian (sacking) are used for rubbing in to achieve the maximum compaction possible. The hessian must have been washed to remove the starch and any impurities. Compaction by rubbing is a very important part of the process and serves to fill the minute hollows of textured stone whilst wiping off all but a smear from the high spots. The treatment is always applied to complete stones, and sometimes, as in the case of sculpture, is carried across joints as well.

Drying out must be as carefully controlled as the drying out of mortar repairs. Polyethylene shrouds are often used and intermittent mist spraying by hand during the first few hours avoids any risk of rapid drying which can result in a powdery and useless shelter coat and undesirable modifications in colour. During the first stages of drying out small additions of colour in the form of finely ground stone dust or even powdered charcoal may be dusted on to achieve minor, subtle variations in the final appearance.

Shelter coating is the most visually striking part of the 'lime method' process but should never be too obvious. Inexpert handling can result in a bland, woolly appearance. However, properly carried out with sufficient sensitivity to the colour and tonal variations of the worn stones it can greatly enhance their appearance.

9 MASONRY CONSOLIDANTS

9.1 INTRODUCTION

This chapter outlines the current use of surface consolidants with particular reference to historic masonry and especially architectural detail. It is not intended as a guide to sculpture conservation, nor does it cover materials and techniques which should properly remain in the province of the museum conservator, although there are areas of common ground.

The idea of a surface treatment which will consolidate friable, delicate stone or ceramic material is obviously very appealing to anyone faced with the problems of conserving either small surface areas of high intrinsic value, or large areas where wholescale replacement seems to be the only other option. The practitioner, unless he or she is fortunate enough to be trained in up-to-date conservation techniques, is often at a loss to know what material, if any, is suitable. A plethora of advertising claims, pseudo-science, prejudices for and against the use of consolidants combine to make decisions and specifications increasingly difficult. After centuries of error and failure in the field there are good reasons for doubting whether there can ever be a practical solution to the problem of significantly slowing down the decay of bricks and stones; at the same time there is a responsibility involved in standing back from the problem. Sometimes, to do nothing is good practice; sometimes to do nothing is culpable neglect. Some appreciation of the potential and limitations of available consolidants is therefore essential.

9.2 WHY USE A CONSOLIDANT?

All masonry materials undergo ageing and deterioration processes. Some alter slowly, in ways which rarely cause alarm and are visually pleasing; others, after a time of slow change, begin to degrade more dramatically, as cementing matrices fail or surface skins spall, blister or scale off as accumulations of soluble salts create internal stresses in the course of cyclic wetting and drying. There is always a simple solution – to replace the distressed material; but, of course, the first aim

Many consolidants are laborious and time consuming to apply. This illustration shows the final stages in the preservation treatment of a Roman marble inset. The weathered marble was filled with a dentistry mortar and then fed with a micro-crystalline wax in a ketones solvent. The warmed surface was finally hand polished to bring back some of the original lustre and colour. Many marbles and dense limestones are notoriously difficult to treat with other consolidants, such as alkoxysilanes.

of the conservator is to conserve. All established and all promising means of extending the life of historic masonry surfaces must, therefore, be of interest.

9.3 WHEN IS THE USE OF A CONSOLIDANT JUSTIFIED?

Broadly speaking, consolidants should only be considered in the following circumstances:

1 When masonry/sculpted surfaces are deteriorating at an obviously unacceptable and readily quantifiable way

2 When the causes of deterioration have been properly identified and adequately understood

3 When there are no practical ways of sufficiently improving the situation by modifying the environment of the victim stones (such as physical moisture barriers, flashings, covers, humidity controls, dentistry and joint repairs, stitching, etc.)

4 When the properties of the proposed consolidant, its constituents and the consolidation technique are known and understood by the specifier and applicator and proper prior consideration has been given to all possible alternative treatments and to laboratory and field performance of the chosen consolidant to date

5 When full records have been made of the subject before treatment and provision has been made to record all steps of the treatment and to establish any necessary periodic maintenance inspections

6 When experienced conservators/applicators are available to carry out the work, bearing in mind that other conservation activities are likely to be involved in the treatment.

9.4 WHICH CONSOLIDANT SHOULD BE USED?

Attempts at consolidating or 'preserving' are by no means new, but the scientific examination of decay mechanisms and 'preservatives' is rather less than one hundred years old.

The most common treatments of the last century, in the UK, effects of which can still be seen and should be recognized, may be listed as follows:

Material	*Appropriate surfaces*
Egg albumen	–
*Lime-casein	Limestone and lime plaster

*Lime-tallow	Limestone and lime plaster
*Limewater (calcium hydroxide)	Limestone and lime plaster
Baryta water (barium hydroxide)	–
Shellac	–
Linseed oil	–
Beeswax	–
Paraffin wax	–
Zinc or magnesium silicofluorides	–
*Potassium and sodium silicates	Lime or cement stucco

The effects of many of these treatments, such as the lime and barium hydroxide systems, will have weathered away; egg and shellac may persist in sheltered areas in the form of lightly discoloured skins with papery flakes; oils and waxes are often heavily discoloured because of their dirt-attracting properties, are still slightly tacky when warm and may be pock-marked by salt crystallization below the thin surface layer; silicofluorides (the only proprietary treatments of the group fall into this category, once known as 'stone liquids' or 'fluate') can survive as unattractive, patchy or streaky grey films accompanied by pitting and flaking.

The lessons of the long and more recent past are clear. Superficial gluing or waterproofing or shallow, pore-blocking treatments are useless or worse than useless, because they exacerbate the condition they are attempting to improve and make better and later treatments difficult or impossible. When stone or ceramic surfaces are decaying, it is not possible to get over the problem with a material, however costly and sophisticated, which modifies only a few millimetres in depth. Superficial treatments should only be applied if they are totally compatible with the surface to which they are applied (which would exclude, for instance, lime-casein applied to polychrome) as in the case of limewater to limestone; or if they are applied as a temporary measure and are genuinely reversible without damage to the treated surface.

Of the traditional treatments, limewater, lime-casein, lime-tallow and barium hydroxide may still claim a useful role on limestone, and potassium and sodium silicate may be used on lime stucco, especially as a vehicle for pigment. This century has added significantly to the consolidant/treatment repertoire. The most useful 'new' materials may be listed as follows:

Material	*Appropriate surfaces*
Alkoxysilanes:	Sandstone and limestone, some ceramics
Ethyl silicate (silicate ester)	
Triethoxymethylsilane	
Trimethoxymethylsilane	
Acrylic polymers:	Sandstone and limestone, some ceramics
Methylmethacrylate	
Acrylics and Siliconesters	

Epoxies	Sandstone and limestone, some ceramics
Polyurethanes	Sandstone and limestone, some ceramics
'Microcrystalline' waxes	Limestone and marbles, some ceramics

These materials, and those traditional treatments asterisked in the list on page 89–90 are likely to remain in current use for many years for the consolidation of sculptured and carved detail on buildings. Some of them, such as the epoxies and polyurethanes have other and better uses than as consolidants. (See Volume 5, Chapter 3, 'The Use of Resin (polymer) Products in the Repair and Conservation of Buildings'). In the external building surface context the alkoxysilane group are undoubtedly the most promising and versatile. Most of them are imported, and although one British system, developed by the Building Research Establishment is probably the best known in the UK, it is amongst the most costly. They are all pore-lining systems with excellent mobility and penetration but have different characteristics suiting them for particular situations.

9.5 PRECAUTIONS

Consolidants, even when carefully and correctly selected, will not perform miracles. They must not be thought of as grouts, void fillers or bridges, or adhesives. They will often be used only as part of a conservation treatment which may include dentistry repair, crack filling, pinning with stainless steel wire and cleaning or poulticing. They should not be applied to zones where there are major moisture movements, to wet surfaces [unless required] or to heavy surface concentrations of soluble salts without some pre-cleaning/poulticing. Suppliers recommendations, in the case of proprietary materials, must be strictly adhered to especially in relation to application method, environmental considerations, protective clothing and respiratory apparatus, protection of work and associated surfaces such as glass. The application of some consolidants is restricted to the supplier's operatives, and the British silane system must be applied by a licensed applicator who carries a card to show that he has been trained to do so. All consolidants, however, should only be applied by experienced applicators who, as a minimum qualification, should be able to discuss the work authoritatively and knowledgeably.

An important part of the research programme of RTAS in English Heritage is the monitoring of consolidant performances; many of the sites still being recorded were treated twelve to fourteen years ago.

10 Colourless Water-Repellent Treatments

General assumptions about the permeability of masonry walls can lead to expensive and unnecessary treatments with water-repellent liquids. In some situations, these treatments can actually increase the incidence of water penetration and, where there is a concentration of soluble salts, may accelerate decay. This section is intended to clarify the role of water-repellent treatments and contains general recommendations relating to their use, based on field experience and the wider recommendations of the Building Research Establishment (see References at the end of this section).

Colourless water repellents are intended to improve the resistance of masonry to rain penetration. Modern water repellents line the pores of the bricks, stones and mortar with water-repellent material, which inhibits capillary absorption. Treated surfaces will still absorb water during prolonged rainfall, but will allow the evaporation of trapped water as the treated zone remains permeable to water vapour.

BRS Digest 125 (January 1971) states that 'water-repellent liquids should be used with discrimination, having regard to the cause of dampness and the suitability of the surface for treatment'. The cause of dampness is often inaccurately diagnosed. If walls are unusually thin, or unusually permeable, water penetration through bricks and stones may take place, especially in conditions of extreme exposure. If penetration persists after all other sensible remedial work has been carried out, such as correct tamping and pointing of joints and cracks and repair of defective copings, gutters, downpipes and flashings, then there may well be a case for the use of water-repellent treatment; but water repellents are not a substitute for other maintenance work. Experience has shown that there are relatively few situations where a water-repellent treatment alone has solved a major damp penetration problem.

The application of water repellents may, in some situations, exacerbate decay. This can happen as a result of water containing salts in solution evaporating from behind the treated layer, leaving salt crystals in the pores behind the treatment. Repeated crystallization cycles can then lead to disruption and spalling of the

This illustration shows the classic situation of decay on the inside of a mullioned window in a marine environment. The outside of the mullion is quite sound, the decay only taking place on the 'drying face' internally. The application of a silicone resin or metallic stearate water repellent to the sound outer face only, is likely to reduce the salt crystallization decay inside. Under no circumstances should these shallow penetration materials treatments be used on decaying stone surfaces, so the inside face must not be treated.

treated surface. In addition to this hazard, the thermal and moisture movements of the thin, treated surface layer may be sufficiently different from those of the underlying stone to generate shear stresses, eventually leading to failure. For these reasons, silicone or other water repellents must not be used, or thought of as consolidants, or 'preservatives' and must not be applied to surfaces which are friable or spalling as the result of salt crystallization damage.

Sometimes, colourless water repellents are applied after cleaning masonry surfaces as dirt inhibitors. Such treatments are successful in this role for the duration of the surface repellency, but this tends to deteriorate relatively quickly, even though the repellence may persist in the surface pores of the treated layer. Retreatment is possible, but is rarely carried out in practice, because of the expense of access to most building facades. Unfortunately, the deterioration of surface repellency is rarely uniform and a patchy appearance can result. Even the short-term benefits of water repellents as dirt inhibitors are, therefore, debatable and rarely justify the costs of materials and labour.

Having advised caution in diagnosis of damp penetration and recommended against the use of water repellents on decayed surfaces, or as dirt inhibitors, there are many other situations where their use does no harm, but is simply an unnecessary expense, perhaps doubling the cost of a small cleaning contract. Any proposal to use them needs, therefore, very careful consideration.

Examples of situations where silicone water repellents have proved their worth must also be acknowledged. The classic situation is masonry close to the sea, where decay is taking place on the inside face of mullions, tracery or lintels, but the outside face is sound. Here, water repellents can usefully be applied to the sound external face only. Elsewhere, frequently on extremely exposed sites, where rain is driven through permeable stones and mortar, water repellents may well justify their use; they have also been used successfully on brickwork where there is a history of staining from limestone dressings, encouraging the calcium carbonate and sulphate to run off the surface rather than be deposited in the bricks.

BS 3826: 1969 'Silicone-based Water Repellents for Masonry' sets out performance standards and offers user guidance as does BS 6477: 1984 'British Standard Specification for Water Repellent Treatments for Masonry Surfaces'. The 1969 standard will continue to operate for three years as the 1984 standard includes a three-year durability test for compliance.

It is advisable to use materials which are manufactured to these standards, or which meet their performance requirements. In BS 6477: 1984 classes of silicone water repellent appropriate for various substrates are as follows:

- *Group 1* Sandstone, clay brick, terracotta, cement and cement-lime stucco
- *Group 2* Limestones and cast stone
- *Group 3* Cement-based materials and other materials of similar alkaline nature
- *Group 4* Calcium-silicate brickwork

When ordering or specifying, the British Standard should be quoted.

REFERENCES

1 Building Research Station, *Building Research Digest 125: Colourless Treatments for Masonry*, HMSO, Garston, January 1971.
2 British Standard 3826, 1969, *Silicone-based Water Repellents for Masonry Surfaces*.
3 British Standard 6477, 1984, *British Standard Specification for Water Repellent Treatments for Masonry Surfaces*.

11 Case study: the consolidation of clunch

11.1 BACKGROUND TO THE PROJECT

During August 1985 a series of comparative consolidation treatments were undertaken on the clunch of the south stable block at Woburn Abbey. The stone at Woburn Abbey had weathered and deteriorated, creating extensive areas of flaking. It had been decided that dressing back was the most satisfactory way of treating this deterioration. As this is a process which can seldom be repeated, requires all stones on a facade to be treated and is relatively labour-intensive and costly, it was decided that the Research and Technical Advisory Service of HBMCE should investigate alternative procedures.

The series of treatments were undertaken on the west elevation of the south stable block. One-half of this elevation had been dressed back and the other had not. The same treatments were applied to both types of surface. The central pedimented bay was not treated. The work was undertaken by the Carvers Conservation Unit of the Research, Technical and Advisory Services of English Heritage.

11.2 THE TREATMENTS APPLIED

The treatments applied were:

- 'Brethane' alkoxy silane consolidant
- Microcrystalline wax
- Boiled linseed oil
- Lime poultice, limewater and lime-casein shelter coat

The location of each treatment was recorded. A control panel was left within both the dressed-back zone and the as found zone. The as found weathered zone was first brushed down with phosphor bronze brushes, treated with a quaternary ammonium biocide, then brushed down again.

Control panels
No 2 (as found), no 9 (dressed back).

Panel 2 received a biocide treatment only.

Brethane panels
No 1 (as found), no 10 (dressed back).

'Brethane' was applied to these panels by trained applicators in full accordance with the specification contained in Ancient Monument Technical Note 15, including recording of treatment. In panel 1 on the untreated side, all loose and flaking material was removed. The edges of all flakes were cut back to sound material using a chisel to form a neat, splayed edge. Both panels were thoroughly brushed down prior to treatment.

Boiled linseed oil panel
No 7 (dressed back).

Boiled linseed oil was applied by brush to a dressed-back panel by the Woburn Abbey mason. It represented the general consolidation approach to the rest of the complex.

Microcrystalline way
No 8 (dressed back).

Panel 8A was treated with pure microcrystalline wax. This was applied liberally and rubbed in with a soft cloth to remove any excess. Panel 8B was treated with a mixture of the wax and pure white spirit. The wax was first dissolved into sufficient white spirit to make a viscous liquid, applied with a brush, then well rubbed in with a soft cloth and any surplus removed.

Lime treatment panels
No 3 (as found), no 6 (dressed back):

- lime poultice, limewater

No 4 (as found), no 5 (dressed back):

- as above plus lime-casein shelter coat

(See Chapter 8, this volume, 'The cleaning and treatment of limestone by the "lime method"' for details of this procedure.)

Lime poultice

A poultice of slaked lump lime was applied to panels 3, 4, 5 and 6. On the as found panels it had a cleaning effect. The panels were first thoroughly dampened, the lime putty applied to a thickness of about 13 mm ($\frac{1}{2}$ in) and then sheeted in plastic to slow down drying. The poultices were removed two days later and the surfaces thoroughly brushed.

Fine jointed clunch ashlar at Woburn Abbey exhibiting a sulphate skin failure visually akin to contour scaling. This kind of disfiguring decay has led to widespread redressing of stone. During the experimental work a case was made for retaining as much of the original face as possible, by limewashing, grouting behind scales and weathering the edges of the scales with fillets of weak lime mortar.

Limewater treatment

Limewater was prepared on site as part of the slaking process. All panels received in excess of 40 applications (300 gallons were applied to panels 5 and 6).

Lime mortar repairs and grouting

Nos 3 and 4.

Panels 3 and 4 had not been dressed back and included several flaking and bulging areas of deterioration. The approach taken here was to consolidate the deteriorated stone without redressing or removing large areas of surface. The edges of flaked pieces of stone were brushed clean, dampened, then supported by lime mortar fillets, formed to a neat, splayed edge. Voids behind such flakes were grouted with a mixture of water, HTI powder, lime putty and acrylic emulsion. Details of the procedures used can be found in Chapter 8, Volume 3, 'Case study: cleaning and consolidation of the chapel plaster at Cowdray House ruins, phases I, II and III'. The ingredients used at Woburn Abbey were based on this experience and were as follows:

Grout

- 1 part hydraulic lime
- $\frac{1}{4}$ part HTI powder
- 3 parts water
- $\frac{1}{10}$ part acrylic emulsion
- $\frac{1}{100}$ part of a 10 per cent solution of sodium gluconate solution

Filleting mix

- 1 part slaked lime putty
- $1\frac{1}{2}$ parts Portland dust
- 1 part silver sand
- Slate dust to colour

Lime-casein shelter coat

Lime-casein shelter coats were applied to panels 4 and 5 on completion of the limewater treatment. For the dressed-back clunch the coat was based on lime putty, Portland dust and yellow staining sand. For the weathered clunch the colour was additionally adjusted by the addition of slate dust. All these ingredients were screened through a 600 micron sieve and mixed with a gauging liquid of half-strength skimmed milk. The coat was made to the consistency of thin cream. Test patches were undertaken and left for two days to dry before confirming the colour match. The shelter coat was brushed well into the dampened surface of the stone and rubbed in hard with a pad of soft hessian, to remove the surplus.

11.3 ASSESSMENT OF TREATMENTS: FIRST INSPECTION

Six weeks after completion of the work, the site was revisited to make an assessment of the various treatments. Similar assessments will be made annually to monitor the weathering of the treatments. The following is a report of the first inspection. The findings were as follows.

Panel 1: Brethane on undressed surface

The Brethane-treated surface was found to have some darkening but this is a temporary, usual effect of this treatment and was considered insignificant when compared to the weathered stone surface adjacent. The surface felt very 'tight'. It was felt that the simple dressing and chamfering of the edges of the sulphated skin was acceptable when coupled with the consolidant in this way.

Panels 3 and 4: lime treatment, undressed surface

The techniques of grouting and filleting were considered very satisfactory in terms of appearance, the very sound surface that was produced and the fact that as much of the original surface was retained as was possible. The lime treatment provided a very sound surface compared to the untreated control.

It was found that weathering had considerably lightened the shelter coat on panel 4, especially in the most heavily weathered section of the panel. It was considered that improvement should be made here.

Panels 5 and 6: lime treatment, dressed surface
These panels were found to be significantly darker than the control. This was contributed to the saturation they had received during the poultice and lime-water treatments. Only a very small amount of dusting was found on the surface. The level of shelter coat on panel 5 was considered good but that an even better colour match could have been achieved. Generally the redressing and joint treatments were accentuated by this treatment.

Panel 7: boiled linseed oil, redressed surface
The darkening of this panel was found to be similar to the Brethane (panel 10). The panel was slightly lighter at the top. Recent treatments of the joints after dressing back were visually accentuated by application of the linseed oil. The surface was exhibiting very little dusting.

Panel 8: microcrystalline wax, redressed surface
Panel 8B, where the wax was applied in white spirit solvent, showed a small amount of dusting. The appearance of this side of the panel was very satisfactory. On panel 8A where the wax was applied full strength the appearance was blotchy and marks from the redressing technique were accentuated.

Panel 9: control, redressed surface
The surface of this panel was found to dust readily.

Panel 10: 'Brethane', redressed surface
The 'Brethane' panel was, predictably, considerably darker and more yellow than the control adjacent. The darkness graded dramatically being lightest at the top. The silane treatment accentuated the redressing marks. No dusting or chalking of the surface was found.

11.4 CONCLUSION AND RECOMMENDATIONS

From a conservation standpoint, the most satisfactory treatment for weathered clunch surface was considered to be limewatering, filleting, grouting and mortar repairs to the as–found surface followed by the application of a shelter coat, coloured to match the weathered colour of the stone and to pick up colour variations.